HSC Year 12
MATHEMATICS STANDARD 2

ADRIAN KRUSE

SERIES EDITOR: ROBERT YEN

A+

2020 UPDATED SYLLABUS · 2020 UPDATED SYLLABUS · 2020 UPDATED SYLLABUS ·

+ topic exams of HSC-style questions
+ practice HSC and mini-HSC exams
+ worked solutions with expert comments
+ HSC exam topic grids (2011–2020)

PRACTICE EXAMS

NELSON
A Cengage Company

A+ HSC Mathematics Standard 2 Practice Exams
1st Edition
Adrian Kruse
ISBN 9780170459211

Publishers: Robert Yen, Kirstie Irwin
Project editor: Tanya Smith
Cover design: Nikita Bansal
Text design: Alba Design
Project designer: Nikita Bansal
Permissions researcher: Corrina Gilbert
Production controller: Karen Young
Typeset by: Nikki M Group

Any URLs contained in this publication were checked for currency during the production process. Note, however, that the publisher cannot vouch for the ongoing currency of URLs.

NSW Education Standards Authority (NESA) 2020 Higher School Certificate Examination Mathematics Standard 1, Mathematics Standard 2 © NSW Education Standards Authority for and on behalf of the Crown in right of the State of New South Wales.

© 2021 Cengage Learning Australia Pty Limited

For product information and technology assistance,
in Australia call **1300 790 853**;
in New Zealand call **0800 449 725**

For permission to use material from this text or product, please email **aust.permissions@cengage.com**

ISBN 978 0 17 045921 1

Cengage Learning Australia
Level 7, 80 Dorcas Street
South Melbourne, Victoria Australia 3205

Cengage Learning New Zealand
Unit 4B Rosedale Office Park
331 Rosedale Road, Albany, North Shore 0632, NZ

For learning solutions, visit **cengage.com.au**

Printed in China by 1010 Printing International Limited.
1 2 3 4 5 6 7 25 24 23 22 21

ABOUT THIS BOOK

Introducing *A+ HSC Year 12 Mathematics*, a new series of study guides designed to help students revise the topics of the new HSC maths courses and achieve success in their exams. *A+* is published by Cengage, the educational publisher of *Maths in Focus* and *New Century Maths*.

For each HSC maths course, Cengage has developed a STUDY NOTES book and a PRACTICE EXAMS book. These study guides have been written by experienced teachers who have taught the new courses, some of whom are involved in HSC exam marking and writing. This is the first study guide series to be published after students sat the first HSC exams of the new courses in 2019–2020, so it incorporates the latest changes to the syllabus and exam format.

This book, *A+ HSC Year 12 Mathematics Standard 2 Practice Exams,* contains topic exams and practice HSC exams, both written and formatted in the style of the HSC exams, with spaces for students to write answers. Worked solutions are provided along with the author's expert comments and advice, including how each exam question is marked. An HSC exam topic grid (2011–2020) guides students to where and how each topic has been tested in past HSC exams.

Mathematics Standard 2 Year 12 topics

1. Linear and non-linear relationships

2. Trigonometry

3. Rates and ratios

4. Investments, loans and annuities

5. Bivariate data and the normal distribution

6. Networks

This book contains:

- 6 topic exams: 1-hour mini-HSC exams on each topic + worked solutions

- 2 practice mini-HSC exams: 1-hour exams + worked solutions

- 2 practice HSC exams: full (2.5-hour) exams + worked solutions

- HSC exam reference sheet of formulas

- bonus: worked solutions to the 2020 HSC exam.

The companion A+ STUDY NOTES book contains topic summaries and graded practice questions, grouped into the same 6 broad topics, including for each topic a concept map, glossary and HSC exam topic grid.

Both books can be used for revision after a topic has been learned as well as for preparation for the trial and HSC exams. Before you begin any questions, make sure you have a thorough understanding of the topic you will be undertaking.

CONTENTS

LINEAR AND NON-LINEAR RELATIONSHIPS

TRIGONOMETRY

RATES AND RATIOS

INVESTMENTS, LOANS AND ANNUITIES

BIVARIATE DATA AND THE NORMAL DISTRIBUTION

YEAR 12 COURSE OVERVIEW

Content for the topic is Mathematics Advanced course.

LINEAR AND NON-LINEAR RELATIONSHIPS

Linear functions

$$y = mx + c,$$
where m is the gradient and c is the y-intercept.

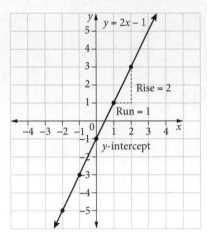

Simultaneous equations

- 2 lines intersect at 1 point.
- Break-even point

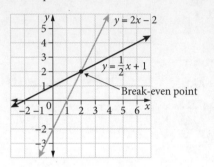

Exponential functions

- Exponential growth: $y = k(a^x)$

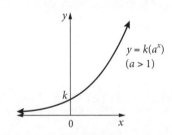

- Exponential decay: $y = k(a^{-x})$

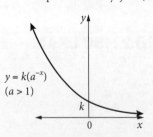

Quadratic functions

$$y = ax^2 + bx + c$$
Its graph is a **parabola**.

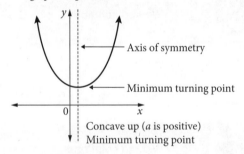

Concave up (a is positive)
Minimum turning point

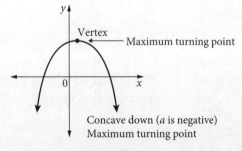

Concave down (a is negative)
Maximum turning point

Direct and inverse variation

- If y is directly proportional to x:
  $$y = kx$$
- If y is inversely proportional to x:
  $$y = \frac{k}{x}$$
- k is the constant of variation.

Reciprocal functions

$$y = \frac{k}{x}$$

Its graph is a **hyperbola**.

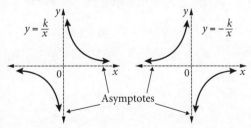

9780170459211

TRIGONOMETRY

Pythagoras' theorem

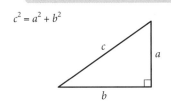

$c^2 = a^2 + b^2$

Right-angled trigonometry

$\sin\theta = \dfrac{\text{opposite}}{\text{hypotenuse}}$ (SOH)

$\cos\theta = \dfrac{\text{adjacent}}{\text{hypotenuse}}$ (CAH)

$\tan\theta = \dfrac{\text{opposite}}{\text{adjacent}}$ (TOA)

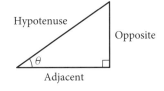

Angles of elevation and depression

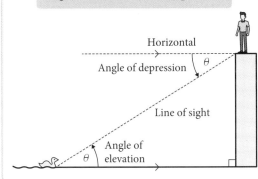

Compass radial surveys

A compass radial survey shows a field with a compass in the middle and the true bearings and distances to each corner.

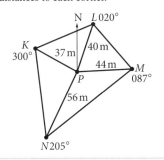

The sine and cosine rules

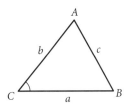

- **Sine rule**

$$\dfrac{a}{\sin A} = \dfrac{b}{\sin B} = \dfrac{c}{\sin C}$$

- **Cosine rule**

$$c^2 = a^2 + b^2 - 2ab\cos C \quad \text{(to find a side)}$$

$$\cos C = \dfrac{a^2 + b^2 - c^2}{2ab} \quad \text{(to find an angle)}$$

Area of a triangle

$$A = \dfrac{1}{2}ab\sin C$$

Bearings

- **Compass bearings:** Acute angle measured from north or south towards east or west. For example, C is N 66° W from O (90° − 24° = 66°).

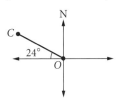

- **True bearings:** 3-digit angles from 000° to 360°, clockwise from north. For example, in the previous diagram, C is 294° from O (270° + 24° = 294°).

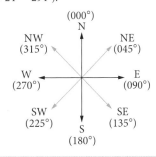

RATES AND RATIOS

Rate problems

- Best buys
- Speed
- Heart rates
- Electricity and energy
- Fuel consumption

Ratio problems

- Simplifying ratios
- Building materials, mixtures and costs
- Dividing a quantity in a given ratio
- Capture–recapture method
- Scales on maps

Scale drawings

- Building plans, elevation views
- Perimeter, area, surface area and volume
- The trapezoidal rule,
$$A \approx \frac{h}{2}(d_f + d_l),$$
to approximate irregular areas

INVESTMENTS, LOANS AND ANNUITIES

Investments

- Simple interest:
$$I = Prn$$
• Compound interest:
future value (FV), present value (PV)
$$FV = PV(1 + r)^n$$
$$I = FV - PV$$
- Comparing simple interest and compound interest
- Inflation and appreciation
- Shares: dividend and dividend yield

$$\text{Dividend yield} = \frac{\text{dividend per share}}{\text{market value per share}} \times 100\%$$

Depreciation

- Straight-line method of depreciation:
$$S = V_0 - Dn$$
- Declining-balance method of depreciation:
$$S = V_0(1 - r)^n$$

Annuities

- Annuities: modelling as a recurrence relation
- Present and future value
- Present and future value tables

Loans and credit cards

- Reducing balance loans
- Credit cards: compound interest

BIVARIATE DATA AND THE NORMAL DISTRIBUTION

Scatterplots

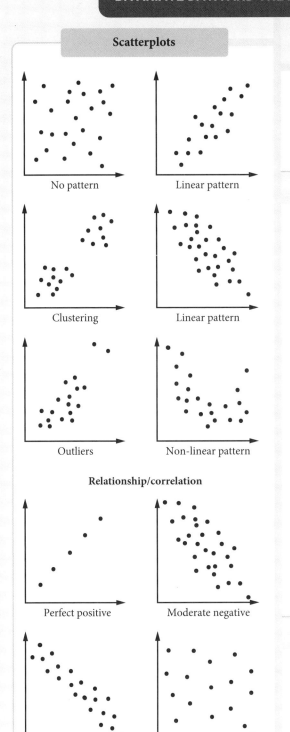

No pattern

Linear pattern

Clustering

Linear pattern

Outliers

Non-linear pattern

Relationship/correlation

Perfect positive

Moderate negative

Strong negative

No relationship

Line of best fit

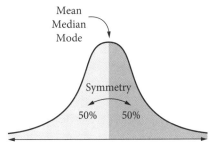

- Line of best fit by eye
- Least-squares regression line by calculator
- Interpolation: data prediction within the set
- Extrapolation: data prediction outside the set

The normal distribution

Mean
Median
Mode

Symmetry

50%   50%

-3 -2 -1 0 1 2 3 z

- Approximately 68% of scores have z-scores between -1 and 1.
- Approximately 95% of scores have z-scores between -2 and 2.
- Approximately 99.7% of scores have z-scores between -3 and 3.

z-scores

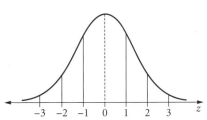

$$z\text{-score} = \frac{\text{score} - \text{mean}}{\text{standard deviation}}$$

$$z = \frac{x - \mu}{\sigma}$$

Pearson's correlation coefficient (r)

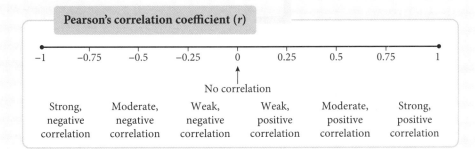

-1 -0.75 -0.5 -0.25 0 0.25 0.5 0.75 1

No correlation

| Strong, negative correlation | Moderate, negative correlation | Weak, negative correlation | Weak, positive correlation | Moderate, positive correlation | Strong, positive correlation |

NETWORKS

Networks terminology

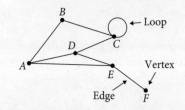

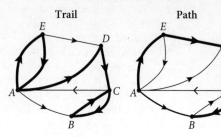

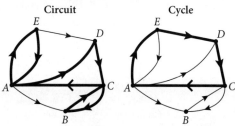

- Directed networks
- Weighted networks
- Connected networks

Minimum spanning trees

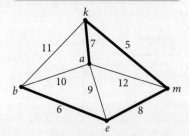

- Kruskal's algorithm
- Prim's algorithm

Shortest path problems

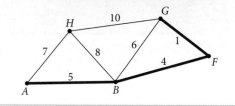

Flow networks

- 'Maximum-flow, minimum-cut' theorem
- Flow capacity

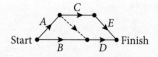

Critical path analysis

- Activity charts and network diagrams

Activity	Predecessor(s)
A	–
B	–
C	A
D	A, B
E	C

- Forward and backward scanning, float times, critical paths

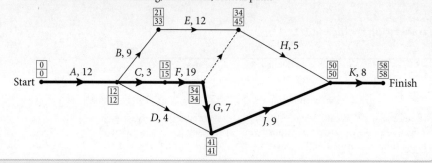

SYLLABUS REFERENCE GRID

Topic and subtopics	A+ HSC Year 12 Mathematics Standard 2 Practice Exams chapter
ALGEBRA	
MS-A4 Types of relationships 　A4.1 Simultaneous linear equations 　A4.2 Non-linear relationships	1 Linear and non-linear relationships
MEASUREMENT	
MS-M6 Non-right-angled trigonometry	2 Trigonometry
MS-M7 Rates and ratios	3 Rates and ratios
FINANCIAL MATHEMATICS	
MS-F4 Investments and loans 　F4.1 Investments 　F4.2 Depreciation and loans	4 Investments, loans and annuities
MS-F5 Annuities	4 Investments, loans and annuities
STATISTICAL ANALYSIS	
MS-S4 Bivariate data analysis	5 Bivariate data and the normal distribution
MS-S5 The normal distribution	5 Bivariate data and the normal distribution
NETWORKS	
MS-N2 Network concepts 　N2.1 Networks 　N2.2 Shortest paths	6 Networks
MS-N3 Critical path analysis	6 Networks

ABOUT THE AUTHOR

Adrian Kruse was Head of Mathematics at Macarthur Anglican School, Cobbity, and has taught for more than 20 years. He has been involved in HSC marking. Adrian is a course specialist for HSC Mathematics Standard 2, creating video tutorials and print content for Edrolo and presenting HSC workshops for The School For Excellence (TSFX). Adrian is currently working at NESA as the 7–10 Mathematics Subject Matter Expert developing new syllabuses.

A+ DIGITAL FLASHCARDS

Revise key terms and concepts online with the A+ Flashcards. Each topic for this course has a deck of digital flashcards you can use to test your understanding and recall. Just scan the QR code or type the URL into your browser to access them.

Note: You will need to create a free NelsonNet account.

https://get.ga/a-hsc-mathsstandard-2

9780170459211

HSC EXAM FORMAT

Mathematics Standard 2 HSC exam

The information below about the Mathematics Standard 2 HSC exam was correct
at the time of printing in 2021. Please check the NESA website in case it has changed.
Visit www.educationstandards.nsw.edu.au, select 'Year 11 – Year 12', 'Syllabuses A–Z',
'Mathematics Standard', then 'Assessment and Reporting in Mathematics Standard Stage 6'.
Scroll down to the heading 'HSC examination specifications for Mathematics Standard 2'.

	Questions	Marks	Recommended time
Section I	15 multiple-choice questions Mark answers on the multiple-choice answer sheet.	15	25 min
Section II	Approx. 25 short-answer questions, including 2 or more questions worth 4 or 5 marks. Write answers on the lines provided on the paper.	85	2 h 5 min
Total		100	2 h 30 min

Exam information and tips

- Reading time: 10 minutes; use this time to preview the whole exam.

- Working time: 2 hours and 30 minutes

- Questions focus on Year 12 outcomes but Year 11 knowledge may be examined.

- Answers are to be written on the question paper.

- A reference sheet is provided at the back of the exam paper, containing formulas.

- Common questions with the Mathematics Advanced HSC exam: 20–25 marks

- Common questions with the Mathematics Standard 1 HSC exam: 20–25 marks

- The 4- or 5-mark questions are usually complex problems that require many steps of working
  and careful planning.

- To help you plan your time, the halfway point of Section II is marked by a notice at the bottom
  of the relevant page; for example, 'Questions 16–27 are worth 44 marks in total'.

- Having 2 hours and 30 minutes for a total of 100 marks means that you have an average
  of 1.5 minutes per mark (or 3 minutes for 2 marks).

- If you budget 20 minutes for Section I and 55 minutes for each half of Section II, then you will have
  20 minutes at the end of the exam to check over your work and complete questions you missed.

STUDY AND EXAM ADVICE

A journey of a thousand miles begins with a single step.

Lao Tzu (c. 570–490 BCE), Chinese philosopher

I've always believed that if you put in the work, the results will come.

Michael Jordan (1963–), American basketball player

Four PRACtical steps for maths study

1. **P**ractise your maths

- Do your homework.

- Learning maths is about mastering a collection of skills.

- You become successful at maths by doing it more, through regular practice and learning.

- Aim to achieve a high level of understanding.

2. **R**ewrite your maths

- Homework and study are not the same thing. Study is your private 'revision' work for strengthening your understanding of a subject.

- Before you begin any questions, make sure you have a thorough understanding of the topic.

- Take ownership of your maths. Rewrite the theory and examples in your own words.

- Summarise each topic to see the 'whole picture' and know it all.

3. **A**ttack your maths

- All maths knowledge is interconnected. If you don't understand one topic fully, then you may have trouble learning another topic.

- Mathematics is not an HSC course you can learn 'by halves' – you have to know it all!

- Fill in any gaps in your mathematical knowledge to see the 'whole picture'.

- Identify your areas of weakness and work on them.

- Spend most of your study time on the topics you find difficult.

4. **C**heck your maths

- After you have mastered a maths skill, such as graphing a quadratic equation, no further learning or reading is needed, just more practice.

- Compared to other subjects, the types of questions asked in maths exams are conventional and predictable.

- Test your understanding with revision exercises, practice papers and past exam papers.

- Develop your exam technique and problem-solving skills.

- Go back to steps 1–3 to improve your study habits.

Topic summaries and concept maps

Summarise each topic when you have completed it, to create useful study notes for revising the course, especially before exams. Use a notebook or folder to list the important ideas, formulas, terminology and skills for each topic. This book is a good study guide, but educational research shows that effective learning takes place when you rewrite learned knowledge in your own words.

A good topic summary runs for 2 to 4 pages. It is a condensed, personalised version of your course notes. This is your interpretation of a topic, so include your own comments, symbols, diagrams, observations and reminders. Highlight important facts using boxes and include a glossary of key words and phrases.

A concept map or mind map is a topic summary in graphic form, with boxes, branches and arrows showing the connections between the main ideas of the topic. This book contains examples of concept maps. The topic name is central to the map, with key concepts or subheadings listing important details and formulas. Concept maps are powerful because they present an overview of a topic on one large sheet of paper. Visual learners absorb and recall information better using concept maps.

When compiling a topic summary, use your class notes, your textbook and the A+ Study Notes book that accompanies this book. Ask your teacher for a copy of the course syllabus or the school's teaching program, which includes the objectives and outcomes of every topic in dot point form.

Attacking your weak areas

Most of your study time should be spent on attacking your weak areas to fill in any gaps in your maths knowledge. Don't spend too much time on work you already know well, unless you need a confidence boost! Ask your teacher, use this book or your textbook to improve the understanding of your weak areas and to practise maths skills. Use your topic summaries for general revision, but spend longer study periods on overcoming any difficulties in your mastery of the course.

Practising with past exam papers

Why is practising with past exam papers such an effective study technique? It allows you to become familiar with the format, style and level of difficulty expected in an HSC exam, as well as the common topic areas tested. Knowing what to expect helps alleviate exam anxiety. Remember, mathematics is a subject in which the exam questions are fairly predictable. The exam writers are not going to ask too many unusual questions. By the time you have worked through many past exam papers, this year's HSC paper won't seem that much different.

Don't throw your old exam papers away. Use them to identify your mistakes and weak areas for further study. Revising topics and then working on mixed questions is a great way to study maths. You might like to complete a past HSC exam paper under timed conditions to improve your exam technique.

Past HSC exam papers are available at the NESA website: visit www.educationstandards.nsw.edu.au and select 'Year 11 – Year 12', 'HSC exam papers'. NESA marking feedback and guidelines can also be viewed there. You can find past HSC exams with solutions online, in bookstores, at the Mathematical Association of NSW (www.mansw.nsw.edu.au) and at your school (ask your teacher) or library.

Preparing for an exam

- Make a study plan early; don't leave it until the last minute.
- Read and revise your topic summaries.
- Work on your weak areas and learn from your mistakes.
- Don't spend too much time studying work you know already.
- Revise by completing revision exercises and past exam papers or assignments.
- Vary the way you study so that you don't become bored: ask someone to quiz you, voice-record your summary, design a poster or concept map, or explain the work to someone.
- Anticipate the exam:
  - How many questions will there be?
  - What are the types of questions: multiple-choice, short-answer, long-answer, problem-solving?
  - Which topics will be tested?
  - How many marks are there in each section?
  - How long is the exam?
  - How much time should I spend on each question/section?
  - Which formulas are on the reference sheet and how do I use them in the exam?

During an exam

1. Bring all of your equipment, including a ruler and calculator (check that your calculator works and is in DEGREES mode for trigonometry). A highlighter pen may help for tables, graphs and diagrams.

2. Don't worry if you feel nervous before an exam – this is normal and helps you to perform better; however, being too casual or too anxious can harm your performance. Just before the exam begins, take deep, slow breaths to reduce any stress.

3. Write clearly and neatly in black or blue pen, not red. Use a pencil only for diagrams and constructions.

4. Use the **reading time** to browse through the exam to see the work that is ahead of you and the marks allocated to each question. Doing this will ensure you won't miss any questions or pages. Note the harder questions and allow more time for working on them. Leave them if you get stuck, and come back to them later.

5. Attempt every question. It is better to do most of every question and score some marks, rather than ignore questions completely and score 0 for them. Don't leave multiple-choice questions unanswered! Even if you guess, you have a chance of being correct.

6. Easier questions are usually at the beginning, with harder ones at the end. Do an easy question first to boost your confidence. Some students like to leave multiple-choice questions until last so that, if they run out of time, they can make quick guesses. However, some multiple-choice questions can be quite difficult.

7. Read each question and identify what needs to be found and what topic/skill it is testing. The number of marks indicates how much time and working out is required. Highlight any important keywords or clues. Do you need to use the answer to the previous part of the question?

8. After reading each question, and before you start writing, spend a few moments planning and thinking.

9. You don't need to be writing all of the time. What you are writing may be wrong and a waste of time. Spend some time considering the best approach.

10. Make sure each answer seems reasonable and realistic, especially if it involves money or measurement.

11. Show all necessary working, write clearly, draw big diagrams, and set your working out neatly. Write solutions to each part underneath the previous step so that your working out goes down the page, not across.

12. Use a ruler to draw (or read) half-page graphs with labels and axes marked, or to measure scale diagrams.

13. Don't spend too much time on one question. Keep an eye on the time.

14. Make sure you have answered the question. Did you remember to round the answer and/or include units? Did you use all of the relevant information given?

15. If a hard question is taking too long, don't get bogged down. If you're getting nowhere, retrace your steps, start again, or skip the question (circle it) and return to it later with a clearer mind.

16. If you make a mistake, cross it out with a neat line. Don't scribble over it completely or use correction fluid or tape (which is time-consuming and messy). You may still score marks for crossed-out work if it is correct, but don't leave multiple answers! Keep track of your answer booklets and ask for more writing paper if needed.

17. Don't cross out or change an answer too quickly. Research shows that often your first answer is the correct one.

18. Don't round your answer in the middle of a calculation. Round at the end only.

19. Be prepared to write words and sentences in your answers, but don't use abbreviations that you've just made up. Use correct terminology and write 1 or 2 sentences for 2 or 3 marks, not mini-essays.

20. If you have time at the end of the exam, double-check your answers, especially for the more difficult or uncertain questions.

Ten exam habits of the best HSC students

1. Has clear and careful working and checks their answers

2. Has a strong understanding of basic algebra and calculation

3. Reads (and answers) the whole question

4. Chooses the simplest and quickest method

5. Checks that their answer makes sense or sounds reasonable

6. Draws big, clear diagrams with details and labels

7. Uses a ruler for drawing, measuring and reading graphs

8. Can explain answers in words when needed, in 1–2 clear sentences

9. Uses the previous parts of a question to solve the next part of the question

10. Rounds answers at the end, not before

Further resources

Visit the NESA website (www.educationstandards.nsw.edu.au) for the following resources.

Select 'Year 11 – Year 12' and then 'Syllabuses A–Z' or 'HSC exam papers'.

- Mathematics Standard 2 Syllabus

- Past HSC exam papers, including marking feedback and guidelines

- Sample HSC questions/exam papers and marking guidelines

Before 2019, 'Mathematics Standard 2' was called 'Mathematics General 2' and, before 2014, 'General Mathematics'. For these exam papers, select 'Year 11 – Year 12', 'Resources archive', 'HSC exam papers archive'.

MATHEMATICAL VERBS

A glossary of 'doing words' common in maths problems and HSC exams

analyse
study in detail the parts of a situation

apply
use knowledge or a procedure in a given situation

calculate
See **evaluate**

classify/identify
state the type, name or feature of an item or situation

comment
express an observation or opinion about a result

compare
show how two or more things are similar or different

complete
fill in detail to make a statement, diagram or table correct or finished

construct
draw an accurate diagram

convert
change from one form to another, for example, from a fraction to a decimal, or from kilograms to grams

decrease
make smaller

describe
state the features of a situation

estimate
make an educated guess for a number, measurement or solution, to find roughly or approximately

evaluate/calculate
find the value of a numerical expression, for example, 3×8^2 or $4x + 1$ when $x = 5$

expand
remove brackets in an algebraic expression, for example, expanding $3(2y + 1)$ gives $6y + 3$

explain
describe why or how

give reasons
show the rules or thinking used when solving a problem. *See also* **justify**

graph
display on a number line, number plane or statistical graph

hence find/prove
calculate an answer or prove a result using previous answers or information supplied

identify
See **classify**

increase
make larger

interpret
find meaning in a mathematical result

justify
give reasons or evidence to support your argument or conclusion. *See also* **give reasons**

measure
determine the size of something, for example, using a ruler to determine the length of a pen

prove
See **show that**

recall
remember and state

show/prove that
(in questions where the answer is given) use calculation, procedure or reasoning to prove that an answer or result is true

simplify
express a result such as a ratio or algebraic expression in its most basic, shortest, neatest form

sketch
draw a rough diagram that shows the general shape or ideas (less accurate than **construct**)

solve
calculate the value(s) of an unknown pronumeral in an equation or inequality

state
See **write**

substitute
replace part of an expression with another, equivalent expression

verify
check that a solution or result is correct, usually by substituting back into an equation or referring back to the problem

write/state
give an answer, formula or result without showing any working or explanation (This usually means that the answer can be found mentally, or in one step)

9780170459211

SYMBOLS AND ABBREVIATIONS

$=$	is equal to	S 37° W	a compass bearing
$\neq$	is not equal to	217°	a true bearing
$\approx$	is approximately equal to	$P(E)$	the probability of event E occurring
$<$	is less than	$P(\bar{E})$	the probability of event E not occurring
$>$	is greater than	LHS	left-hand side
$\leq$	is less than or equal to	RHS	right-hand side
$\geq$	is greater than or equal to	%	percentage
( )	parentheses, round brackets	p.a.	per annum (per year)
[ ]	(square) brackets	cos	cosine ratio
{ }	braces	sin	sine ratio
$\pm$	plus or minus	tan	tangent ratio
π	pi = 3.141 59...	$\bar{x}$	the mean
$0.\dot{1}5\dot{2}$	the recurring decimal 0.152 152...	σ_n	the standard deviation
°	degree	Σ	the sum of, sigma
$\angle$	angle	Q_1	first quartile or lower quartile
Δ	triangle	Q_2	median (second quartile)
$\therefore$	therefore	Q_3	third quartile or upper quartile
x^2	x squared, $x \times x$	IQR	interquartile range
x^3	x cubed, $x \times x \times x$	α	alpha
$\sqrt{}$	square root	θ	theta
$\sqrt[3]{}$	cube root	μ	micro-, mu, population mean
		m	gradient

A+ HSC YEAR 12 MATHEMATICS

STUDY NOTES

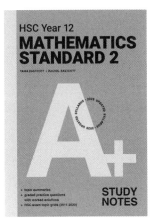   

Authors:
Tania Eastcott
Rachel Eastcott

Sarah Hamper

Karen Man
Ashleigh Della Marta

Jim Green
Janet Hunter

PRACTICE EXAMS

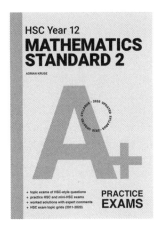   

Authors:
Adrian Kruse

Simon Meli

John Drake

Jim Green
Janet Hunter

9780170459211

CHAPTER 1
TOPIC EXAM

Linear and non-linear relationships

MS-A4 Types of relationships

    A4.1 Simultaneous linear equations

    A4.2 Non-linear relationships

- A reference sheet is provided on page 200 at the back of this book
- For questions in Section II, show relevant mathematical reasoning and/or calculations

Reading time: 4 minutes
Working time: 1 hour
Total marks: 40

Section I – 6 questions, 6 marks
- Attempt Questions 1–6
- Allow about 10 minutes for this section

Section II – 11 questions, 34 marks
- Attempt Questions 7–17
- Allow about 50 minutes for this section

Section I

- Attempt Questions 1–6
- Allow about 10 minutes for this section

6 marks

Question 1

Which of the graphs below best represents $y = x^2 - 3$?

A

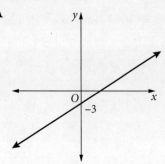

B

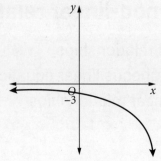

C

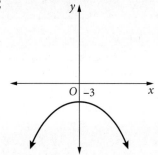

D

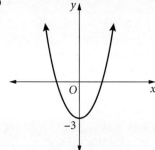

Question 2

Graphs of the equations $y = x + 2$ and $y = -x + 4$ are shown below.

What is the solution when the equations are solved simultaneously?

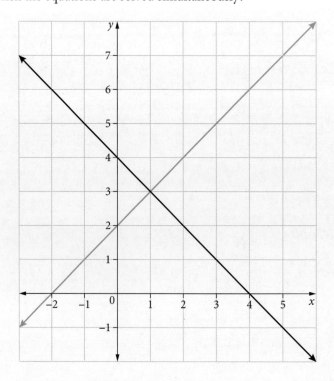

A $y = 1, x = 3$ **B** $x = 1, y = 3$ **C** $x = 1, y = 1$ **D** $x = 3, y = 3$

Question 3

What is the equation for the graph below?

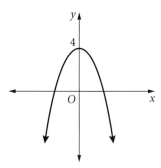

A $y = x^2 + 4$ **B** $y = -x^2 + 4$

C $y = x + 4$ **D** $y = x^4$

Question 4

Which of the graphs below represents $y = 2^{-x}$?

A

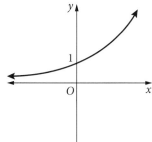

B

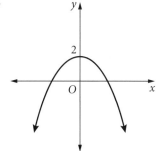

C

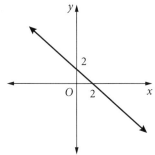

D

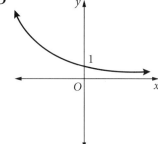

Question 5

Which of the graphs below best represents $y = \dfrac{2}{x}$ for positive values of x?

A

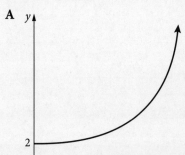

B

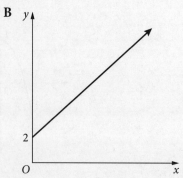

C

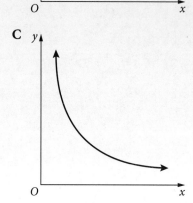

D

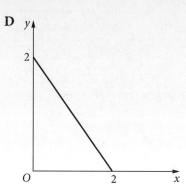

Question 6

The graph below shows an exponential decrease of 30% each year.

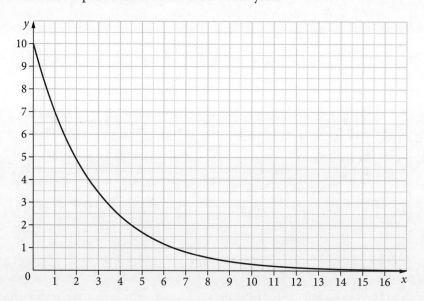

Which of the following equations is correct for the graph shown?

A $y = 10(0.7)^x$ **B** $y = (0.7)^x$

C $y = 10(0.3)^x$ **D** $y = (0.3)^x$

Section II

• Attempt Questions 7–17	**34 marks**
• Allow about 50 minutes for this section	
• Answer the questions in the spaces provided. These spaces provide guidance for the expected length of response.	
• Your responses should include relevant mathematical reasoning and/or calculations.	

Question 7 (2 marks)

On the number plane below, graph the lines $y = 2x - 6$ and $y = \frac{1}{2}x$ and then solve this pair of equations simultaneously.   2 marks

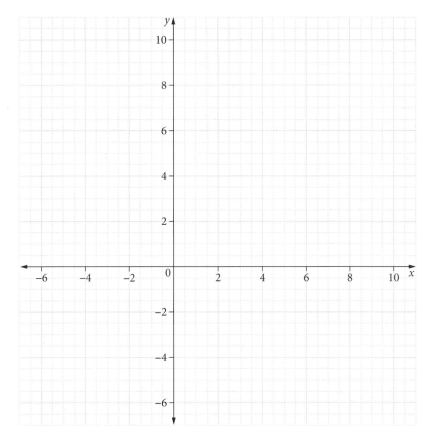

_____

_____

_____

Question 8 (3 marks)

The city of Peisville experienced growth in population according to the function
$P = 2000(1.08)^x$, where P is the population and x is the number of years after 1996.
The graph of the function is shown below.

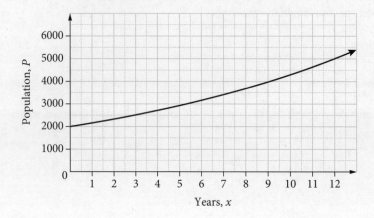

a What was the population of the town in 1996?   1 mark

_____

b At what rate is the population of the town growing each year?   1 mark

_____

_____

c In what year was the population approximately twice the size of the population in 1996?   1 mark

_____

_____

Question 9 (3 marks)

The graph of the equation $y = \dfrac{2}{x}$ is shown below.

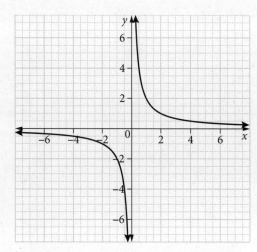

a For what value(s) of x is the value of y not defined?   1 mark

_____

b What are the asymptotes of this graph? 1 mark

c What would be the effect on the graph if $y = \dfrac{2}{x}$ was changed to $y = \dfrac{8}{x}$? 1 mark

Question 10 (4 marks)

The graph below shows the height of an egg that has been thrown into the air that then returns to the ground.

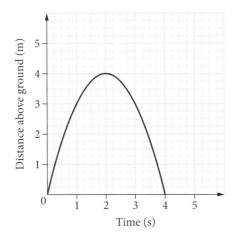

a What was the maximum height of the egg above the ground? 1 mark

b For how long was the egg in the air? 1 mark

c At what time(s) was the egg 3 metres above the ground? 1 mark

d Why does the graph not extend after 4 seconds on the horizontal axis? 1 mark

Question 11 (2 marks)

A group of 10 friends go camping and they take enough food to last 14 days. The number of people going camping (N) is inversely proportional to the number of days of food supply (F).

If 4 extra people join the group on the first day (who don't bring extra food), how long will the food supply last?

2 marks

Question 12 (2 marks)

This is the graph of a reciprocal function.

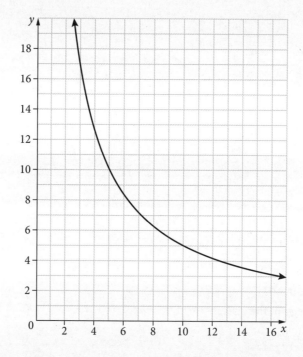

What is the equation of the curve?

2 marks

Question 13 (3 marks)

Joe decides to host a party for some of his friends. He hires a catering company, which charges a $500 set-up fee plus a charge of $15 for every person who attends. To cover his costs, Joe charges each of his friends $40 to attend the party.

a Write a formula for the cost (C) in dollars of holding the party for x number of people.   1 mark

b The graph below shows the planned income and costs for the party.

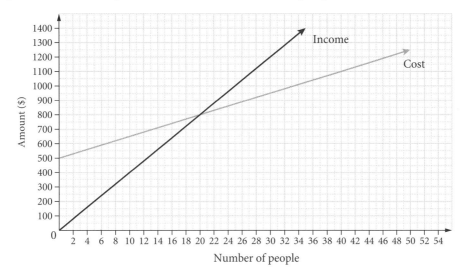

How many paying guests will need to attend in order for Joe to break even?   1 mark

c How much profit will Joe make if 30 people attend the party?   1 mark

Questions 7–13 are worth 19 marks in total (Section II halfway point)

Question 14 (4 marks)

Nadine has 60 m of fencing. She wants to put the entire length around a rectangular garden. The graph below shows the possible widths of the garden and the different areas of the garden for each width.

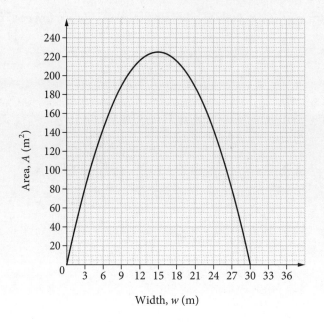

a   If the perimeter of the garden is found using $P = 2l + 2w$, show that $l = 30 - w$.   1 mark

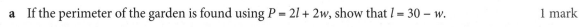

b   Hence, show that the area of the rectangular garden is given by $A = 30w - w^2$.   1 mark

c   Using the graph or another method, find the maximum possible area of the garden.   1 mark

d   Why doesn't the graph exist for $w < 0$ and $w > 30$?   1 mark

Question 15 (3 marks)

The speed of Jarrod's internet connection, I, in megabits per second (Mbps) is inversely proportional to the number of users (U) in the local area. When there are 2000 users, the speed is 24 Mbps.

How many of the 2000 users would need to go offline for the speed to increase to 96 Mbps?   3 marks

Question 16 (4 marks)

A scientist has two groups of different bacteria. Each group of bacteria grows according to a different exponential model.

Bacteria A grow at an exponential rate with the formula $N = 5000(1.26)^x$ and Bacteria B grow at a rate with the formula $N = 5000(1.12)^x$. In these models, N is equal to the number of bacteria and x is the number of days elapsed. The growth according to both models is shown below.

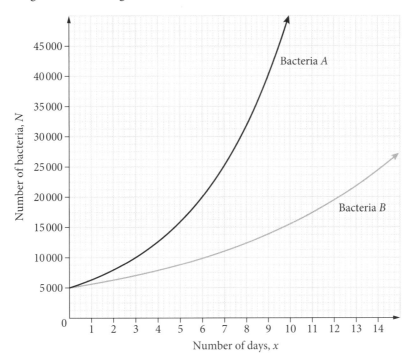

a What was the initial number of bacteria for both groups?   1 mark

b What was the difference in growth rates between the two groups of bacteria?   1 mark
Express your answer as a percentage.

c What was the difference in bacteria populations after 6 days?   1 mark

d How many more days did it take for Bacteria B to reach a population of 10 000   1 mark
than for Bacteria A?

Question 17 (4 marks)

A construction company wants to employ a certain number of workers to complete a project. The more workers it employs, the less time it will take to complete the project. The relationship between the number of workers and completion time of the project is modelled by the equation $W = \dfrac{200}{n}$, where W is the number of workers employed and n is the number of years it will take to complete the project.

a Use the equation given to fill the blank cells in the following table.   1 mark

n	0.5	1	2	3	4	6
W				67	50	33

b Use the table of values to draw the graph of the relationship below.   2 marks

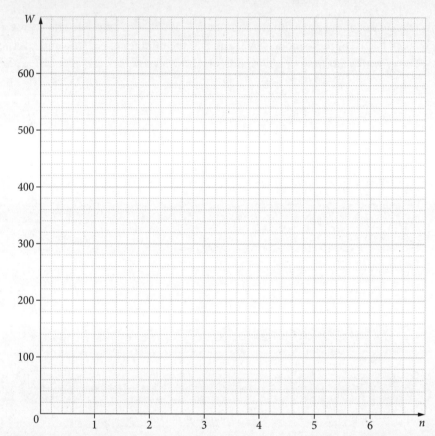

c Give one limitation of the model.   1 mark

_____

_____

_____

END OF PAPER

WORKED SOLUTIONS

Section I (1 mark each)

Question 1

D A 'concave up' parabola with y-intercept -3.

> Understand the shape of a quadratic function and how to manipulate the curve.

Question 2

B The point of intersection is always the solution of two straight lines.

> Remember that the solution is always an x- and y-value.

Question 3

B A negative sign in front of the x^2 will reflect the parabola in the x-axis and the '+ 4' will translate the curve up 4 units.

> Straightforward question. Know how to manipulate non-linear functions.

Question 4

D A negative power in an exponential will reflect the curve in the y-axis.

> A question that tests your knowledge of manipulating an exponential function.

Question 5

C An equation in the form $y = \dfrac{k}{x}$, with x on the denominator of a fraction, is a reciprocal function.

> Some HSC questions only give you a picture of the first quadrant. Be mindful of this and know what that curve would look like in all four quadrants as well as in a single quadrant.

Question 6

A A 30% decrease means that you are left with 70% of the original value, hence the 0.70 in the brackets. The curve intersects the y-axis at 10, so when $x = 0$, $y = 10$.

> Be familiar with the initial amount in an exponential equation as this is where the curve will cut the y-axis. Also, pay attention to the percentage in the brackets.

Section II (✓ = 1 mark)

Question 7 (2 marks)

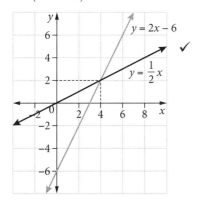

$x = 4$, $y = 2$ or $(4, 2)$ ✓

> Always use a ruler and a table of values to graph each line. Remember to write *both* the x- and y-coordinates when giving the solution of two equations.

Question 8 (3 marks)

a 2000 ✓

As shown by the equation and graph, the initial population of the town was 2000.

b 8% ✓

The number in the brackets (1.08) represents the initial amount being multiplied by 108%, so the rate at which the town is growing each year is 8%.

c 2005 ✓

Looking at the graph, a population of approximately 4000 occurs 9 years after 1996, that is, in 2005.

> Be prepared to read information from a variety of non-linear graphs and always read the question carefully to make sure your answer makes sense.

Question 9 (3 marks)

a $x = 0$ ✓

Division by 0, $\frac{2}{0}$, is not defined.

b x- and y-axes, where $x = 0$ and $y = 0$. ✓

c The curves would be further away from $(0, 0)$. There is a vertical dilation by a factor of 4. ✓

> Be familiar with the different parts of any non-linear function and know how to manipulate each function.

Question 10 (4 marks)

a Maximum height = 4 m ✓

> Look at the y-value of the turning point of the graph.

b The egg was in the air for 4 seconds. ✓

> The graph exists between $x = 0$ and $x = 4$ on the x-axis.

c

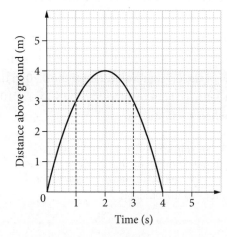

The egg was 3 metres above the ground at $t = 1$ and $t = 3$ seconds. ✓

> This is an example of a quadratic function: there are 2 times (solutions) where the egg is at a height of 3 metres above the ground – once on the way up and once on the way down. There are 2 times for each height of the egg, except for at the turning point.

d The graph does not continue past 4 on the horizontal axis (t-axis) because there is no negative distance (distance below the ground) in this question. ✓

> Part **d** is a common question asking about limitations of the model. Be prepared to write worded answers with logical reasoning in the context of the question.

Question 11 (2 marks)

$N = \dfrac{k}{F}$ $N = 10, F = 14$

$10 = \dfrac{k}{14}$ So $k = 140$ ✓

Therefore, $N = \dfrac{140}{F}$

$N = 10 + 4 = 14$

$14 = \dfrac{140}{F}$ So $F = \dfrac{140}{14}$

$= 10$ days ✓

> Most inverse variation questions are structured in the same way. These are usually straightforward questions. Remember to find the value of k first.

Question 12 (2 marks)

$y = \dfrac{k}{x}$

From the graph, choose $x = 10$ and $y = 5$: ✓

$5 = \dfrac{k}{10}$

So $k = 50$

So $y = \dfrac{50}{x}$ ✓

> Look for clear-cut points on the graph like $(10, 5)$ that will help you find the value of k, hence allowing you to find the inverse variation equation.

Question 13 (3 marks)

a $C = 15x + 500$ ✓

Fixed cost is always on its own and variable cost is always in front of the variable.

b The break-even point is 20 people. ✓

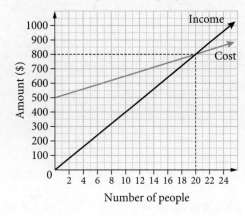

c Profit = income − expenses

$$= \$1200 - \$950$$

$$= \$250 \checkmark$$

> Break-even questions are common in HSC exams. Understand the meaning of the point of intersection between a cost and income (or revenue) graph.

Question 14 (4 marks)

a $P = 2l + 2w$

$60 = 2l + 2w$

$2l = 60 - 2w$

$l = \dfrac{60}{2} - \dfrac{2w}{2}$

$ = 30 - w \checkmark$

b $A = l \times w$

$ = (30 - w) \times w \checkmark$

$ = 30w - w^2$

> Practise making formulas from a worded problem. Questions that link quadratic equations to area have been asked a number of times in HSC exams.

c Maximum area is the turning point of the graph, $(15, 225)$, so area = $225\,\text{m}^2$.

OR using the formula, when $w = 15$:

maximum area = $30(15) - 15^2$

$\phantom{\text{maximum area}} = 225\,\text{m}^2 \checkmark$

d The graph does not exist for $w < 0$ because you cannot have negative widths or areas. The graph does not exist for $w > 30$ because you cannot have negative areas, and for 60 m of fencing it is not possible to have a width of more than $\frac{1}{2} \times 60 = 30\,\text{m}$. $\checkmark$

Question 15 (3 marks)

$I = \dfrac{k}{U} \qquad I = 24, U = 2000$

So $24 = \dfrac{k}{2000}$

$k = 48\,000$

So $I = \dfrac{48\,000}{U}$ $\checkmark$

$96 = \dfrac{48\,000}{U}$

So $U = \dfrac{48\,000}{96}$

$ = 500 \checkmark$

So $2000 - 500 = 1500$. $\checkmark$

So 1500 users would need to go offline so that the speed would increase to 96 Mbps.

> Note that the question asks how many users would need to go *offline*, not how many users are online. Always make sure that you have answered the question in the correct context.

Question 16 (4 marks)

a The initial number of bacteria in each group is 5000. $\checkmark$

b Bacteria A = 26% Bacteria B = 12%

Therefore, difference is 14%. $\checkmark$

> Remember to look at the numbers in the brackets, for example, (1.26) means 126% which is a 26% increase.

c

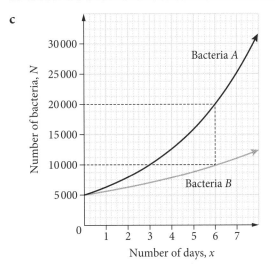

Difference = $20\,000 - 10\,000$

$ = 10\,000$

The difference in bacteria populations after 6 days is 10 000. $\checkmark$

d Bacteria A reached 10 000 in 3 days. Bacteria B reached 10 000 in 6 days.

Therefore, Bacteria B took 3 days longer to reach 10 000 than Bacteria A. $\checkmark$

> Be familiar with percentages and reading information from both axes of an exponential graph. This is an important skill in this topic.

Question 17 (4 marks)

a
n	0.5	1	2	3	4	6
W	400	200	100	67	50	33
✓

b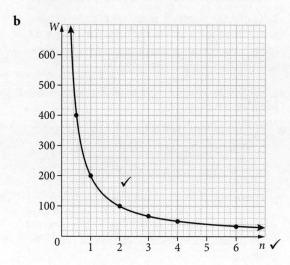
✓

c There will be a limit as to how many workers the company can hire. Also, if there are too many workers, then they will get in each other's way and slow the project down. ✓

> Be prepared to graph from a table of values and comment in words on the practical limitations of an algebraic model. These types of questions have been asked in previous HSC exam papers.

HSC exam topic grid (2011–2020)

This grid shows the coverage of this topic in past HSC exams by question number. The past exams can be downloaded from the NESA website (www.educationstandards.nsw.edu.au) by selecting 'Year 11 – Year 12', 'HSC exam papers'. NESA marking feedback and guidelines can also be found there.

Before 2019, 'Mathematics Standard 2' was called 'Mathematics General 2' and, before 2014, 'General Mathematics'. For these exam papers, select 'Year 11 – Year 12', 'Resources archive', 'HSC exam papers archive'.

	Algebra and equations (Year 11)	Linear functions	Non-linear functions	Direct and inverse variation
2011	12, 18	20*, 23(b), 28(b)	6[#], 26(b)[#]	28(a)
2012	14, 21, 28(b)	5, 8, 13	16, 30(b)–(c)[#]	15
2013	5, 21, 29(a)	28(b)	22, 30(a)	
2014	4, 11, 26(a), 26(c), 29(b)	7, 26(d)[*†], 29(a)	3	26(f)
2015	2, 23, 24, 26(b), 28(d)	13, 28(f)[*]	10[#], 29(e)	
2016	2, 5, 24, 26(b)	4[*], 14, 29(e)	29(b)[#]	
2017	7, 9, 19, 20, 28(a)(i), 28(d), 30(d)(i)	17[*†]	28(e)	
2018	16, 25, 26(b), 28(b), 28(e)	27(d)	4	29(c)
2019 new course	11, 28	14, 23(c), 36[*]	31	33, 34
2020	13	6, 10, 24[*]	1, 19, 33[#]	

* Simultaneous equations.
† Can be solved graphically.
[#] Exponential function.

CHAPTER 2
TOPIC EXAM

Trigonometry

MS-M6 Non-right-angled trigonometry

- A reference sheet is provided on page 200 at the back of this book
- For questions in Section II, show relevant mathematical reasoning and/or calculations

Reading time: 4 minutes
Working time: 1 hour
Total marks: 40

Section I – 6 questions, 6 marks
- Attempt Questions 1–6
- Allow about 10 minutes for this section

Section II – 15 questions, 34 marks
- Attempt Questions 7–21
- Allow about 50 minutes for this section

Section I

- Attempt Questions 1–6   **6 marks**
- Allow about 10 minutes for this section

Question 1

Find the value of x in the triangle, correct to the nearest metre.

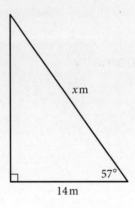

A 24   **B** 26   **C** 58   **D** 59

Question 2

Find the area of the triangle, correct to two decimal places.

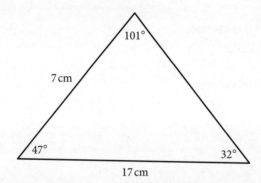

A $31.53\,\text{cm}^2$ **B** $43.52\,\text{cm}^2$ **C** $53.45\,\text{cm}^2$ **D** $58.40\,\text{cm}^2$

Question 3

What is the true bearing of G from H?

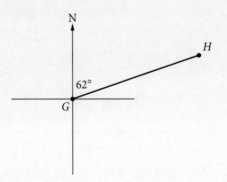

A $062°$ **B** $118°$ **C** $172°$ **D** $242°$

Question 4

What is the length of *AB* in the triangle below, correct to one decimal place?

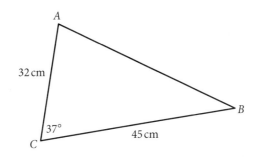

A 27.4 cm   **B** 35.1 cm   **C** 47.9 cm   **D** 748.9 cm

Question 5

Romer goes to visit Julia, who lives at the top of a tower. Julia spots Romer at an angle of depression of 32°. Romer is 21 m from the base of the tower. Julia drops a rope for Romer to climb up and see her.

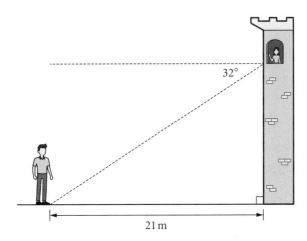

How far does Romer have to climb vertically up the rope to reach Julia's window?

A 13.1 m   **B** 14.1 m   **C** 15.1 m   **D** 16.1 m

Question 6

What is the size of the largest angle in the triangle, to the nearest minute?

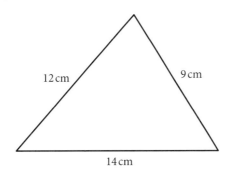

A 82°17′   **B** 82°18′   **C** 89°55′   **D** 89°56′

Section II

- Attempt Questions 7–21
- Allow about 50 minutes for this section
- Answer the questions in the spaces provided. These spaces provide guidance for the expected length of response.
- Your responses should include relevant mathematical reasoning and/or calculations.

34 marks

Question 7 (1 mark)

Write the correct expression for $\tan 33°$ for this triangle.   1 mark

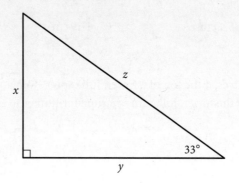

Question 8 (1 mark)

Find the value of θ in this triangle, correct to the nearest minute.   1 mark

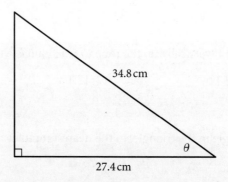

Question 9 (2 marks)

Find the distance from *F* to *G*, correct to one decimal place.   2 marks

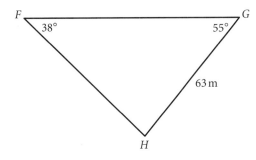

Question 10 (2 marks)

In the triangle below, *FG* = 45 cm, *FH* = 53 cm and *GH* = 28 cm.

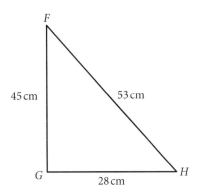

a   Prove that Δ*FGH* is a right-angled triangle.   1 mark

b   Hence, calculate the size of ∠*FHG*, correct to the nearest minute.   1 mark

Question 11 (2 marks)

Find the value of x, correct to the nearest whole number, if the area of this triangle is $522\,\text{m}^2$.   2 marks

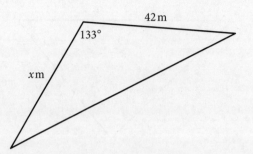

Question 12 (3 marks)

Dane performed a radial survey of his backyard and drew the following diagram.

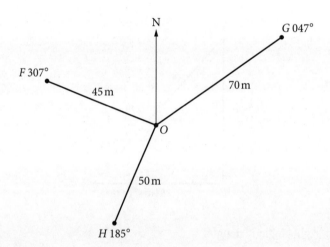

a What is the size of $\angle FOG$?   1 mark

b What is the shortest distance between F and H, correct to the nearest metre?   2 marks

Question 13 (2 marks)

Three hot air balloons are all flying at the same height. The diagram represents a view of the balloons from above.

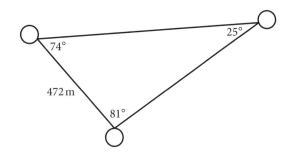

What is the longest distance between two balloons, correct to the nearest metre?   2 marks

_____

_____

_____

_____

Question 14 (3 marks)

Alison has a garden bed that is a perfect circle with a diameter of 10 m. She wants to set up a concrete triangle inside the circle with one vertex at the centre of the circle. The rest of the garden bed will contain soil or grass.

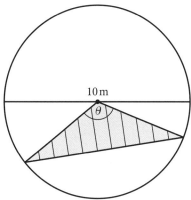

a Show that the total area of the circular garden bed is 78.5 m^2.   1 mark

_____

_____

b The local council has told Alison that her planned concrete triangle must be 15% of the total garden area.

If Alison is to comply with this regulation, what is angle θ for the triangle? Give your answer to the nearest degree.   2 marks

_____

_____

_____

_____

Question 15 (1 mark)

A ramp connects a footpath to a road. According to the local council's safety guidelines, the ramp must have a slope that is in the ratio of $1:8$. This means there is a rise of 1 unit for every 8 units of horizontal distance.

What is the angle of inclination of the ramp (to the horizontal), correct to the nearest minute?   1 mark

_____

_____

_____

_____

Questions 7–15 are worth 17 marks in total (Section II halfway point)

Question 16 (2 marks)

In the diagram, U is north-west of S and $\angle USV = 158°$.

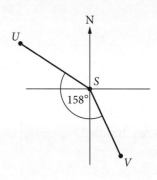

What is the bearing of S from V?   2 marks

_____

_____

_____

Question 17 (3 marks)

In the diagram, $AD = AC$.

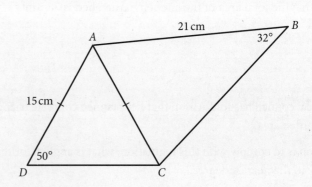

What is the size of $\angle DCB$, correct to the nearest minute?   3 marks

_____

_____

_____

_____

Question 18 (4 marks)

Sue stands on top of a 35 m high cliff. She spots a sailboat out at sea, approaching the base of the cliff. She first sees the boat when it is at an angle of depression of 11°5′. She sees the same boat 30 seconds later, now at an angle of depression of 16°23′.

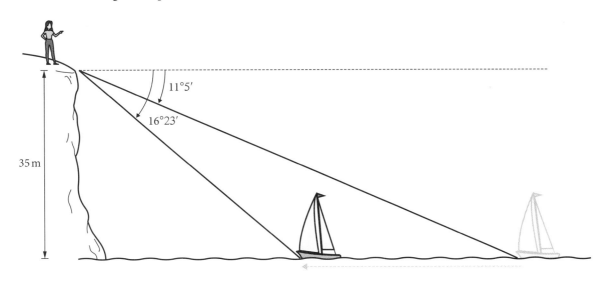

a How far has the boat travelled in 30 seconds, correct to the nearest metre?   3 marks

b What is the speed of the boat, in metres per second?   1 mark

Question 19 (3 marks)

ΔDEF is an acute-angled triangle.

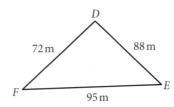

Find the area of ΔDEF, correct to the nearest square metre.   3 marks

Question 20 (3 marks)

Phil is walking home and 34 m from his destination when flash flooding blocks his path. He has to take a detour around the flooding.

He turns 57° and walks for 24 m. He then walks directly home.

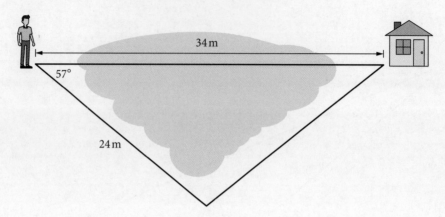

How much further did Phil have to walk via the detour compared to his direct route home?   3 marks
Give your answer correct to two decimal places.

Question 21 (2 marks)

B is on a bearing of 340° from *U*.

B is 14 km from *U* and 16 km from *G*.

U is 12 km from *G*.

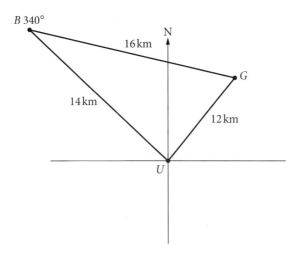

What is the bearing of *G* from *B*, correct to the nearest minute?   2 marks

_____

_____

_____

_____

_____

_____

_____

_____

_____

_____

END OF PAPER

WORKED SOLUTIONS

Section I (1 mark each)

Question 1

B $\cos\theta = \dfrac{\text{adj}}{\text{hyp}}$

$\cos 57° = \dfrac{14}{x}$

$x = \dfrac{14}{\cos 57°}$

$= 25.705...$

$\approx 26\,\text{m}$

> Straightforward question. If a triangle has a right angle, use Pythagoras' theorem (when given 3 sides) or SOH CAH TOA (when given 2 sides and an angle).

Question 2

B $A = \dfrac{1}{2}ab\sin C$

$= \dfrac{1}{2} \times 7 \times 17 \times \sin 47°$

$= 43.5255...$

$\approx 43.52\,\text{cm}^2$

> Make sure that the angle that you use is between the 2 sides that are being substituted into the formula.

Question 3

D

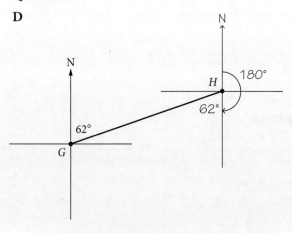

At H, there is an alternate angle to the 62° at G between the 2 parallel lines pointing north, so the angle at H is also 62°.

From the diagram, the bearing of G from H is $180° + 62° = 242°$.

> Drawing a compass rose on the point after the word 'from' (H) helps with this type of question.

Question 4

A $c^2 = a^2 + b^2 - 2ab\cos C$

$AB^2 = 32^2 + 45^2 - 2(32)(45)\cos 37°$

$AB = \sqrt{748.929...}$

$\approx 27.4\,\text{cm}$

> The cosine rule is used when given 3 sides or 2 sides and 1 angle. Remember that the unknown side length must be opposite the angle used in the formula.

Question 5

A

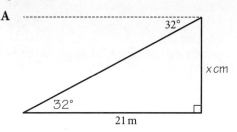

$\tan\theta = \dfrac{\text{opp}}{\text{adj}}$

$\tan 32° = \dfrac{x}{21}$

$x = 21 \times \tan 32°$

$\approx 13.1\,\text{m}$

> Remember that the angle of depression is outside the triangle that you are using. Use properties of alternate angles on parallel lines to find the interior angle.

Question 6

A $\cos C = \dfrac{a^2 + b^2 - c^2}{2ac}$

$\cos\theta = \dfrac{12^2 + 9^2 - 14^2}{2 \times 12 \times 9}$

$= 0.1342...$

$\theta = \cos^{-1} 0.1342...$

$\approx 82°17'$

> The largest angle is always opposite the longest side.

Section II ($\checkmark$ = 1 mark)

Question 7 (1 mark)

$$\tan\theta = \frac{\text{opp}}{\text{adj}}$$

$$\tan 33° = \frac{x}{y}$$

Therefore, $\dfrac{x}{y}$ is the correct expression. $\checkmark$

Always label the sides of the triangle after locating your reference angle.

Question 8 (1 mark)

$$\cos\theta = \frac{\text{adj}}{\text{hyp}}$$

$$= \frac{27.4}{34.8}$$

$$\theta = 38°3'39''$$

$$\approx 38°4' \ \checkmark$$

Make sure you read the question carefully all the way until the end to make sure that you round your answers correctly. A question may include rounding as part of obtaining a mark.

Question 9 (2 marks)

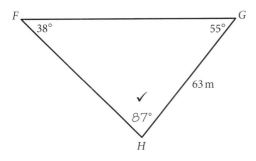

$$\frac{FG}{\sin 87°} = \frac{63}{\sin 38°}$$

$$FG = \frac{63}{\sin 38°} \times \sin 87°$$

$$\approx 102.2\,\text{m} \ \checkmark$$

Make sure that you have opposite sides and angles in each fraction of the sine rule.

Question 10 (2 marks)

a $FH^2 = 53^2 = 2809$ (Pythagoras' theorem)

$$FG^2 + GH^2 = 45^2 + 28^2$$

$$= 2809$$

$$= FH^2 \ \checkmark$$

Therefore, ΔFGH is a right-angled triangle since Pythagoras' theorem holds true.

b $\sin\angle FHG = \dfrac{\text{opp}}{\text{hyp}}$

$$= \frac{45}{53}$$

So $\angle FHG = 58°7'$. $\checkmark$

If the first part of a question requires you to 'prove' a result, remember that if you cannot prove the result, you can still use the result to answer the other parts of the question.

Question 11 (2 marks)

$$A = \frac{1}{2}ab\sin C$$

$$522 = \frac{1}{2} \times 42 \times x\sin 133°$$

$$x = \frac{522 \times 2}{42\sin 133°} \ \checkmark$$

$$= 33.98\ldots$$

$$\approx 34\,\text{m} \ \checkmark$$

Make sure that the angle is *between* the 2 sides given when finding the area of a triangle.

Question 12 (3 marks)

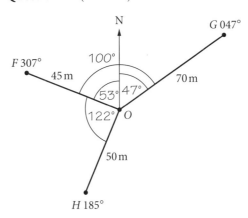

a $\angle FOG = 360° - 307° + 47°$

$$= 100° \ \checkmark$$

b $\angle FOH = 307° - 185°$

$$= 122° \ \checkmark$$

$$c^2 = a^2 + b^2 - 2ab\cos C$$

$$FH^2 = 45^2 + 50^2 - 2(45)(50)\cos 122°$$

$$FH = \sqrt{6909.63\ldots}$$

$$\approx 83\,\text{m} \ \checkmark$$

When an angle turns through north, make sure you add the angles either side of the 0°/360° line.

Question 13 (2 marks)

Let x = longest side opposite the angle 81°.

$$\frac{x}{\sin 81°} = \frac{472}{\sin 25°}$$

$$x = \frac{472}{\sin 25°} \times \sin 81° \quad ✓$$

$$\approx 1103 \text{ m} \quad ✓$$

The longest side is always opposite the largest angle.

Question 14 (3 marks)

a $A = \pi r^2 \qquad r = \frac{10}{2} = 5 \text{ m}$

$= \pi(5)^2$

$\approx 78.5 \text{ m}^2 \quad ✓$

b $78.5 \times 0.15 = 11.775 \text{ m}^2 \quad ✓$

$$A = \frac{1}{2}ab\sin C$$

$$11.775 = \frac{1}{2} \times 5 \times 5 \times \sin\theta$$

$$\sin\theta = \frac{11.775}{12.5}$$

$$\theta \approx 70° \quad ✓$$

It is quite common for questions to require you to rearrange equations to make the unknown variable the subject. Make sure you practise these types of questions.

Question 15 (1 mark)

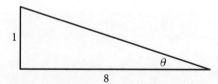

$$\tan\theta = \frac{\text{opp}}{\text{adj}}$$

$$= \frac{1}{8}$$

$$\theta = 7°8' \quad ✓$$

It is always helpful to draw a diagram if it is not given.

Question 16 (2 marks)

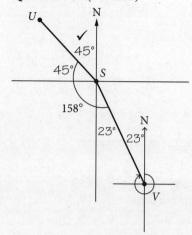

Angle between SV and south = 158° − 45° − 90°

$= 23°$

Bearing of S from V = 360° − 23° = 337° ✓

Always draw a compass rose on the point after the word 'from'.

Question 17 (3 marks)

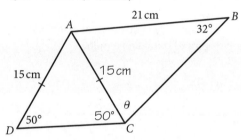

$\angle ADC = \angle ACD$

So $\angle ACD = 50°$ ✓

So $AC = AD = 15 \text{ cm}$

$$\frac{\sin\theta}{21} = \frac{\sin 32°}{15}$$

$$\sin\theta = 0.7418\ldots$$

$$\theta \approx 47°54' \quad ✓$$

So $\angle DCB = 50° + 47°54'$

$= 97°54' \quad ✓$

Make sure that you are familiar with basic geometrical principles such as properties of an isosceles triangle.

Question 18 (4 marks)

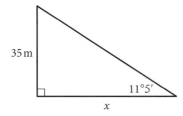

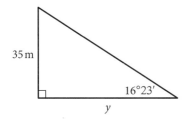

a $\tan 11°5' = \dfrac{35}{x}$

$x = \dfrac{35}{\tan 11°5'}$

$= 178.671\ldots$ ✓

$\tan 16°23' = \dfrac{35}{y}$

$y = \dfrac{35}{\tan 16°23'}$ ✓

$= 119.047\ldots$

$\approx 119\,\text{m}$

So distance travelled $= 178.671\ldots - 119.047\ldots$

$= 59.6245\ldots$

$= 60\,\text{m}$ ✓

b $\text{Speed} = \dfrac{\text{distance}}{\text{time}}$

$= \dfrac{60}{30}$

$= 2\,\text{m/s}$ ✓

Drawing the 2 triangles will help with answering this question.

Question 19 (3 marks)

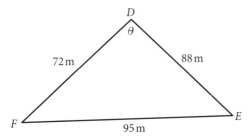

$\cos C = \dfrac{a^2 + b^2 - c^2}{2ac}$

$\cos \theta = \dfrac{72^2 + 88^2 - 95^2}{2(72)(88)}$

$= 0.308\ldots$

$\theta \approx 72°4'$ ✓

Area of $\Delta DEF = \dfrac{1}{2}ab\sin C$

$A = \dfrac{1}{2} \times 72 \times 88 \times \sin 72°4'$ ✓

$= 3013.99\ldots$

$\approx 3014\,\text{m}^2$ ✓

To find the area of a triangle, you only need to find 1 angle.

Question 20 (3 marks)

Let unknown side $= x$

$c^2 = a^2 + b^2 - 2ab\cos C$

$x^2 = 24^2 + 34^2 - 2(24)(34)\cos 57°$

$x = \sqrt{843.149\ldots}$

$= 29.037\ldots$ ✓

Therefore, total distance of the detour

$= 24 + 29.037$

$\approx 53.04\,\text{m}$ ✓

So extra distance $= 53.04 - 34$

$= 19.04\,\text{m}$ ✓

Always make sure you determine what the question is asking and answer appropriately, including the correct units.

Question 21 (2 marks)

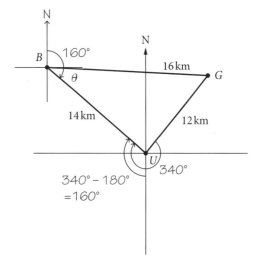

$\cos \theta = \dfrac{14^2 + 16^2 - 12^2}{2(14)(16)}$

$= 0.6875\ldots$

$\theta \approx 46°34'$ ✓

$\angle NBU = 160°$ (alternate angle with $160°$ at U)

So bearing of G from $B = 160° - 46°34'$

$= 113°26'$ ✓

Drawing a compass rose on each point will help obtain the bearings that are required.

HSC exam topic grid (2011–2020)

This table shows the coverage of this topic in past HSC exams by question number. The past exams can be downloaded from the NESA website (www.educationstandards.nsw.edu.au) by selecting 'Year 11–Year 12', 'HSC exam papers'. NESA marking feedback and guidelines can also be found there.

Before 2019, 'Mathematics Standard 2' was called 'Mathematics General 2', and before 2014, 'General Mathematics'. For these exams, select 'Year 11–Year 12', 'Resources archive', 'HSC exam papers archive'.

	Right-angled trigonometry	The sine and cosine rules	Area of a triangle	Bearings
2011	4, 9	24(c)		24(c)
2012	4, 27(d)	20, 29(c)	10	20
2013	4	24, 26(a)	28(a)*	28(a)*
2014	26(b)	28(b)*	28(b)*	23, 28(b)*
2015	9	30(e)	22	7*
2016	26(d)	25, 30(c)	30(c)	25
2017	8, 26(d)	30(c)		30(c)
2018	22, 30(c)	12	30(c)	7
2019 new course	12, 22	17	35*	4, 35*
2020	16	31, 32	32	31

* Radial survey.

CHAPTER 3
TOPIC EXAM

Rates and ratios

MS-M7 Rates and ratios

• A reference sheet is provided on page 200 at the back of this book • For questions in Section II, show relevant mathematical reasoning and/or calculations	**Reading time: 4 minutes** **Working time: 1 hour** **Total marks: 40**

Section I – 6 questions, 6 marks
• Attempt Questions 1–6
• Allow about 10 minutes for this section

Section II – 13 questions, 34 marks
• Attempt Questions 7–19
• Allow about 50 minutes for this section

Section I

· Attempt Questions 1–6	**6 marks**
· Allow about 10 minutes for this section	

Question 1

Zoe's heart beats 22 times in 15 seconds.

What is Zoe's heart rate in beats per minute (bpm)?

A 44 bpm   **B** 86 bpm

C 88 bpm   **D** 330 bpm

Question 2

What is the simplest form of the ratio 4.5 : 17?

A 9 : 34   **B** 2.25 : 8.5

C 18 : 68   **D** 34 : 9

Question 3

A car is travelling at 75 km/h.

What is the car's speed in metres per second?

A 18.4 m/s   **B** 20.8 m/s

C 23.1 m/s   **D** 24.7 m/s

Question 4

A map has a scale of 1 : 200 000. The distance between two campsites on the map is 4.7 cm.

What is the actual distance, in kilometres?

A 2 km   **B** 7 km

C 9.4 km   **D** 43.2 km

Question 5

In a pod of 520 dolphins swimming off the coast, 180 are male.

What is the ratio of male to female dolphins, expressed as a simplified ratio?

A 9 : 17   **B** 17 : 9

C 180 : 340   **D** 340 : 180

Question 6

A car has a fuel tank capacity of 60 L. On a recent trip with a full tank of petrol, the car travelled 420 km. At the end of the trip, the fuel gauge indicated that the tank was 20% of its capacity.

What is the average fuel consumption rate of the car?

A 4.8 L/100 km   **B** 10.6 L/100 km

C 11.4 L/100 km   **D** 48 L/100 km

Section II

• Attempt Questions 7–19	**34 marks**
• Allow about 50 minutes for this section	
• Answer the questions in the spaces provided. These spaces provide guidance for the expected length of response.	
• Your responses should include relevant mathematical reasoning and/or calculations.	

Question 7 (1 mark)

Cement, sand and gravel are mixed in the ratio of $1:2:3$ to make concrete.

A landscaper has 30 kg of gravel.

How much sand will he need to make concrete?   1 mark

Question 8 (2 marks)

A 680 g box of cornflour costs \$5.70, while a 300 g box costs \$3.20.

What is the difference in price per kilogram between these two box sizes?   2 marks

Question 9 (3 marks)

Orchid drives her car to and from work 5 days a week. The distance between her home
and work is 61 km. Her car has an average fuel consumption rate of 6.6 L/100 km.

If petrol costs an average of \$1.30/L for the year and Orchid works for 48 weeks of the year,   3 marks
how much is her yearly petrol bill for travelling to and from work?

Question 10 (3 marks)

A sports bar installs 18 LED TV screens that each have a power of 30 watts. The screens are all on for 16 hours a day, every day of the year.

If electricity costs $0.28/kWh and the sports bar is eligible for a 15% discount on their electricity bill, how much do the screens cost to run for a year?   3 marks

_____

_____

_____

_____

_____

Question 11 (2 marks)

Lauren is a Shar Pei dog breeder. She has 14 puppies that are either black or tan in colour in the ratio of $5:2$. Ten more puppies join the litter.

How many of the 10 new puppies must be tan in colour to change the ratio to $5:3$?   2 marks

_____

_____

_____

_____

_____

_____

_____

_____

Question 12 (1 mark)

Three brothers bought 126 rolls of toilet paper in the ratio $8:2:4$.

What was the largest number of toilet rolls purchased by 1 brother?   1 mark

_____

_____

_____

_____

Question 13 (3 marks)

Indie's target heart rate (THR) is calculated using the following formula:

$$THR = I \times (MHR - RHR) + RHR,$$

where I is the exercise intensity, MHR is the maximum heart rate and RHR is Indie's resting heart rate.

Her MHR is calculated using the following formula:

$$MHR = 226 - \text{her age}$$

Indie achieves a THR of 160 with a workout intensity of 0.75. If she has a resting heart rate of 70 beats per minute, how old is she?   3 marks

Question 14 (2 marks)

Richard is studying rabbit populations in an outback region. On his first trip, he captures 300 rabbits, tags them and then releases them. Two weeks later, he comes back and captures 200 of them and notices that only 30 of them have tags. He estimates that, in the next 2 years, the rabbit population will increase by 14.2%.

What is Richard's estimate of the rabbit population 2 years from now?   2 marks

Questions 7–14 are worth 17 marks in total (Section II halfway point)

9780170459211

Question 15 (3 marks)

Jack and Jill participate in a 6-minute yoga routine. Jack's heart beats 420 times, whereas Jill's heart beats 480 times for the duration of the routine.

Calculate the difference between Jack's and Jill's heart rates, as a percentage of Jack's heart rate, correct to one decimal place.

3 marks

Question 16 (2 marks)

Below is a map of roads out of Sydney.

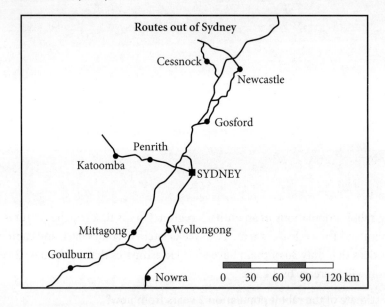

a If the scale key is 4 cm long, what is the scale ratio?

1 mark

b If the distance from Gosford to Newcastle is 90 km, what is the distance on the map between the two cities?

1 mark

TOPIC EXAM

Question 17 (4 marks)

Terry is restoring the roof on the apartment block shown on the map below. (The map is not drawn to scale.)

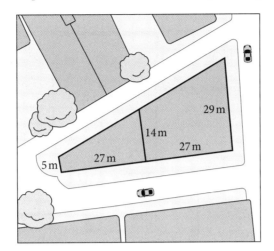

a   Use two applications of the trapezoidal rule to approximate the area of the roof.   1 mark

b   Terry designs the roof so that the rainwater can be collected into storage tanks below.

How many litres of water will be collected if 50 mm of rain falls?   2 marks

c   If the scale for the map was 1 : 500, what length on the map would be equal to a real   1 mark
length of 29 m?

Question 18 (5 marks)

This is the ground floor plan of a split-level house.

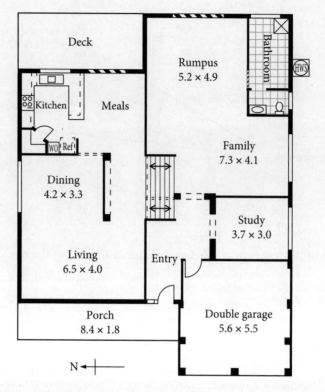

a How many hinged doors are indicated on the plan?   1 mark

b Show that the area of the double garage is approximately 2.8 times the area of the study.   1 mark

c The living and dining areas are to be tiled. Each tile is 900 mm × 900 mm. The tiles are available in boxes of 5 and tilers must order 10% extra for cutting and wastage.

How many boxes of tiles need to be purchased?   3 marks

Question 19 (3 marks)

Farmer Jane purchases the piece of farmland pictured below.

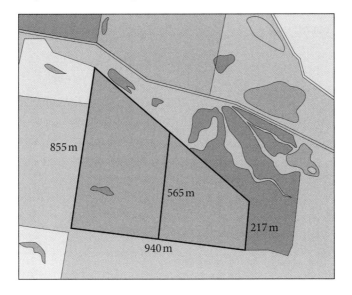

a   Use the trapezoidal rule to show that the area of the property is $517\,470\,\text{m}^2$.   1 mark

b   During the rainy season, a total volume of $30\,000\,\text{kL}$ of rain fell on the property.

How much rain (in mm) fell on the property? Answer to the nearest millimetre.   2 marks

END OF PAPER

WORKED SOLUTIONS

Section I   (1 mark each)

Question 1

C 15 seconds : 22 times

60 seconds : 15 seconds × 4, so

Zoe's heart rate = 22 × 4

$\qquad\qquad$ = 88 bpm

> Always look at the units required in the answer. In this case they are beats per minute.

Question 2

A $4.5 : 12 = 2 \times 4.5 : 2 \times 17$

$\qquad\qquad = 9 : 34$

> Remember that ratios in simplest form are whole numbers.

Question 3

B Speed $= \dfrac{75\,\text{km/h}}{3.6}$

$\qquad \approx 20.8\,\text{m/s}$

OR Speed = 75 km/h

$\qquad\quad$ = 75 000 m/h

$\qquad\quad = \dfrac{75000}{3600}\,\text{m/s}$ $\qquad$ (1 h = 3600 s)

$\qquad\quad \approx 20.8\,\text{m/s}$

> The quickest way to convert km/h to m/s is to divide by 3.6.

Question 4

C Scale = 1 : 200 000

So 4.7 cm represents 200 000 cm × 4.7

= 940 000 cm

= 9400 m $\qquad\qquad$ (1 m = 100 cm)

= 9.4 km $\qquad\qquad$ (1 km = 1000 m)

> Remember that in a scale the units are the same.

Question 5

A Female = 520 − 180

$\qquad\quad$ = 340

So male : female = 180 : 340

$\qquad\qquad\qquad$ = 9 : 17

> Note the order of items in the question because order is important in ratios.

Question 6

C Petrol used = 80% × 60

$\qquad\qquad$ = 48 L

Fuel consumption = 4.8 L/420 km

$\qquad\qquad\quad = \dfrac{4.8\,\text{L}}{4.2} \bigg/ \dfrac{420\,\text{km}}{4.2}$

$\qquad\qquad\quad \approx 11.4\,\text{L/100 km}$

> With fuel consumption questions, make sure to have '/100 km' in the rate.

Section II ($\checkmark = 1$ mark)

Question 7 (1 mark)

Gravel, 3 parts = 30 kg

So 1 part = 10 kg

Sand, 2 parts = 20 kg $\checkmark$

> Straightforward question. Always try to find out what one part is equal to (unit) for 1 quantity.

Question 8 (2 marks)

Price per kg (large box) $= \dfrac{\$5.70}{0.68\,\text{kg}}$

$\approx \$8.38$ $\checkmark$

Price per kg (small box) $= \dfrac{\$3.20}{0.3\,\text{kg}}$

$\approx \$10.67$

Difference = $10.67 - $8.38

$= \$2.29$ $\checkmark$

> Comparing best buys always involves converting all prices to a common unit. Choose a unit that either the question asks for or something that is an appropriate unit for comparison.

Question 9 (3 marks)

Total distance $= 61 \times 2 \times 5 \times 48$

$= 29\,280\,\text{km}$ $\checkmark$

Petrol used $= \dfrac{29\,280}{100} \times 6.6$

$= 1932.48\,\text{L}$ $\checkmark$

Petrol cost = 1932.48 × $1.30

$= \$2512.22$ $\checkmark$

> These marks are easily attainable if the question is approached systematically and logically. Look out for questions that include 'return' trips.

Question 10 (3 marks)

$30\,\text{W} = \dfrac{30}{1000}\,\text{kW} = 0.03\,\text{kW}$

Power for 1 year $= 0.03 \times 18 \times 16 \times 365$

$= 3153.6\,\text{kWh}$ $\checkmark$

Electricity cost = 3153.6 kWh × $0.28/kWh

$= \$883.01$ $\checkmark$

Cost with 15% discount = $883.01 × 0.85

$= \$750.56$ $\checkmark$

> Remember that the power usage of an appliance is the number of kilowatts it draws after 1 hour of continuous use, measured in kWh. Electricity costs are calculated in kWh.

Question 11 (2 marks)

Total parts = 5 + 2 = 7

So 7 parts = 14 puppies

So 1 part = 2 puppies

So there are 5 × 2 = 10 black, 2 × 2 = 4 tan $\checkmark$

New ratio, total parts = 5 + 3

Total parts = 8

8 parts = 24 puppies

1 part = 3 puppies

So there are 5 × 3 = 15 black, 3 × 3 = 9 tan

Therefore, she needs 9 − 4 = 5 more tan puppies to change the ratio to 5 : 3. $\checkmark$

> After working out what 1 part will equal in the new ratio, you can easily obtain the answer.

Question 12 (1 mark)

Total parts = 8 + 2 + 4 = 14

14 parts = 126 rolls

So 1 part = 126 ÷ 14 = 9 rolls

Largest number = 8 parts

So 8 × 9 = 72 rolls $\checkmark$

> Always find out what 1 part is equal to.

Question 13 (3 marks)

$$\text{THR} = I \times (\text{MHR} - \text{RHR}) + \text{RHR}$$
$$160 = 0.75 \times (\text{MHR} - 70) + 70$$
$$160 - 70 = 0.75 \times (\text{MHR} - 70) \checkmark$$
$$\frac{90}{0.75} = \text{MHR} - 70$$
$$120 + 70 = \text{MHR}$$
$$\text{MHR} = 190 \checkmark$$

$$\text{MHR} = 226 - \text{age}$$
$$190 = 226 - \text{age}$$
$$\text{age} = 226 - 190$$

So Indie's age = 36 $\checkmark$

A lot of these formulas require algebraic manipulation. Make sure you practise how to rearrange formulas to find different variables.

Question 14 (2 marks)

$$\frac{\text{total}}{\text{tagged 1st visit}} = \frac{\text{total in sample}}{\text{tagged in sample}}$$
$$\frac{x}{300} = \frac{200}{30}$$
$$x = \frac{200}{30} \times 300$$
$$= 2000 \text{ rabbits} \checkmark$$

2 years from now = 2000 × 1.142 (14.2% increase)
= 2284 rabbits $\checkmark$

Capture–recapture problems are all very similar and can be easily solved. There are different methods of solving them. Choose the method that works for you and always structure your answer in the same way.

Question 15 (3 marks)

$$\text{Jack} = \frac{420}{6}$$
$$= 70 \text{ beats per minute}$$

$$\text{Jill} = \frac{480}{6} \checkmark$$
$$= 80 \text{ beats per minute}$$

Difference = 80 − 70
= 10 $\checkmark$

Difference as % of Jack's pulse = $\frac{10}{70} \times 100$
$$= 14.3\% \checkmark$$

This is a straightforward question with a small twist at the end. Percentages are common twists at the end of questions.

Question 16 (2 marks)

a 4 cm : 120 km = 4 cm : 12 000 000 cm
= 1 : 3 000 000 $\checkmark$

b 4 cm : 120 km
So 1 cm is 30 km.
For a distance of 90 km:
Scaled distance = $\frac{90}{30}$ = 3 cm
So 90 km is 3 cm on the map. $\checkmark$

When working with scales, always ensure you convert the measurements so you have the same units on both sides.

Question 17 (4 marks)

a $A \approx \frac{27}{2}(5 + 14) + \frac{27}{2}(14 + 29)$
$$\approx 837 \text{ m}^2 \checkmark$$

b $V = Ah$
$= 837 \times 0.05$ $\quad (50 \text{ mm} = \frac{50}{1000} = 0.05 \text{ m})$
$= 41.85 \text{ m}^3 \checkmark$
$= 41.85 \text{ kL}$
$= 41 850 \text{ L} \checkmark$

c 1 cm is 500 cm
= 5 m
So a scaled length for 29 m is $\frac{29}{5}$ = 5.8 cm. $\checkmark$

Rainfall questions inevitably lead to volume questions. Remember that rain falling on a roof can be approximated by a prism, which you can then find the volume of using $V = Ah$ (where h is the amount of rainfall).

Question 18 (5 marks)

a 3 hinged doors $\checkmark$

b Area of garage = 5.6 × 5.5
= 30.8 m^2

Area of study = 3.7 × 3.0
= 11.1 m^2

Area comparison = $\frac{30.8}{11.1}$
$$\approx 2.8 \checkmark$$

c Area of living and dining

$= (4.2 \times 3.3) + (6.5 \times 4.0)$

$= 39.86\,\text{m}^2$

Area of 1 tile $= 0.9\,\text{m} \times 0.9\,\text{m}$

$= 0.81\,\text{m}^2$ ✓

Number of tiles required $= \dfrac{39.86}{0.81}$

$= 49.2$

Extra 10% $= 49.2 \times 1.1$

$= 54.1$ ✓

Number of boxes $= \dfrac{54.1}{5}$

$= 10.824$

So number of boxes required $= 11$ ✓

Round up for number of boxes. Be familiar with symbols on house plans as well as working with measurements on a plan.

Question 19 (3 marks)

a $\dfrac{1}{2} \times 940\,\text{m} = 470\,\text{m}$

$A \approx \dfrac{470}{2}(855 + 565) + \dfrac{470}{2}(565 + 217)$

$\approx 517\,470\,\text{m}^2$ ✓

b $V = Ah$ (Remember that $1\,\text{m}^3 = 1\,\text{kL}$.)

$30\,000 = 517\,470h$ ✓

$h = \dfrac{30\,000}{517\,470}$

$= 0.057\,97\ldots\,\text{m}$

$\approx 58\,\text{mm}$ ✓

Remember that rainfall on a piece of land can be approximated by a prism. The height of the prism is the rainfall in mm.

HSC exam topic grid (2011–2020)

This grid shows the coverage of this topic in past HSC exams by question number. The past exam papers can be downloaded from the NESA website (www.educationstandards.nsw.edu.au) by selecting 'Year 11 – Year 12', 'HSC exam papers'. NESA marking feedback and guidelines can also be found there.

Before 2019, 'Mathematics Standard 2' was called 'Mathematics General 2' and, before 2014, 'General Mathematics'. For these exams, select 'Year 11 – Year 12', 'Resources archive', 'HSC exam papers archive'.

	Rates	Ratios	Scale and scale drawings	Trapezoidal rule (introduced 2019)
2011	21		24(a)	13[†]
2012	26(g)	26(f)*	27(c)	
2013	26(d)	30(c)		
2014	17, 18, 20, 22, 27(a)(i), 27(b)		28(d)	28(d)(iii)[†]
2015	26(b), 26(g), 27(b), 30(a), 30(d)	26(a)*	29(c)	
2016	9, 11, 15, 26(c), 28(b)	28(a)*		
2017	2, 6, 14, 26(a), (b), 27(d)(iii)	26(c)*		
2018	5, 8, 26(g), 27(a), 27(d)(i), 28(c)	10*	26(g)	28(a)[†]
2019 new course	2, 24, 33(a), 41(a)	18	41(a)	41(b)
2020	3	23	27	27

* Capture–recapture method.

† Uses Simpson's rule, which is no longer in the course, but can be solved by the trapezoidal rule, giving a similar answer.

CHAPTER 4
TOPIC EXAM

4

Investments, loans and annuities

MS-F4 Investments and loans
    F4.1 Investments
    F4.2 Depreciation and loans
MS-F5 Annuities

• A reference sheet is provided on page 200 at the back of this book • For questions in Section II, show relevant mathematical reasoning and/or calculations	**Reading time: 4 minutes** **Working time: 1 hour** **Total marks: 40**

Section I – 6 questions, 6 marks
• Attempt Questions 1–6
• Allow about 10 minutes for this section

Section II – 12 questions, 34 marks
• Attempt Questions 7–18
• Allow about 50 minutes for this section

9780170459211

Section I

•   Attempt Questions 1–6   **6 marks**   •   Allow about 10 minutes for this section

TOPIC EXAM

Question 1

What is the future value of $17 000 invested at 4.9% p.a. compounded every 6 months for 9 years?

A $21 998.00   **B** $24 458.90   **C** $26 282.35   **D** $27 352.35

Question 2

In 1994, $100 was equivalent in purchasing power to $188.28 in 2020.

What was the average rate of inflation over the 26 years from 1994 to 2020?

A 2.25%   **B** 2.46%   **C** 3.33%   **D** 3.46%

Question 3

Alan owns 3220 shares with a total market value of $101 591.00.

The dividend he receives per share is $2.80.

Calculate the dividend yield.

A 0.002%   **B** 7.2%   **C** 8.9%   **D** $31.55

Question 4

A motorbike is purchased for $22 000. The value of the motorbike is depreciated by 12.2% each year using the declining-balance method.

By how much does the value of the motorbike decrease in 3 years?

A $7109.60   **B** $8052.00   **C** $12 448.90   **D** $14 890.40

Question 5

Kloe uses her credit card to purchase a new doghouse for $380. Her credit card has a no interest-free period and interest charges of 16% p.a. compounded daily on balances outstanding.

If this is the only transaction on the credit card and she pays her balance in full after 20 days, how much interest is she charged for the transaction?

A $2.86   **B** $3.35   **C** $382.86   **D** $383.35

Question 6

John borrows $200 000 and is charged 8% p.a. interest compounded monthly.

If the first repayment of $2030 is made at the end of the first month, what is the opening balance at the beginning of the second month?

A $196 777.67   **B** $197 970.00   **C** $198 666.67   **D** $199 303.33

Section II

> - Attempt Questions 7–18   **34 marks**
> - Allow about 50 minutes for this section
> - Answer the questions in the spaces provided. These spaces provide guidance for the expected length of response.
> - Your responses should include relevant mathematical reasoning and/or calculations.

Question 7 (2 marks)

Samuel buys 800 shares in company XYZ. The current market price is $3.80 per share and pays a dividend of $0.36 per share.

a What is Samuel's total dividend?   1 mark

b What is the dividend yield on Samuel's shares, correct to one decimal place?   1 mark

Question 8 (2 marks)

Kayoum purchased a big-screen LED TV, which depreciated in value by 18% each year. After 3 years, the TV was valued at $4686.63 using the declining-balance method.

What was the purchase price of Kayoum's TV?   2 marks

Question 9 (2 marks)

Breana is charged compound interest at a rate of 0.044 38% per day on any credit card balance that is outstanding. There is a 30-day interest-free period on her credit card, but if payment occurs after this period, then interest is calculated from the date of purchase. Breana has $985 outstanding on her card for airline tickets bought 52 days ago.

How much interest is Breana charged?   2 marks

TOPIC EXAM

Question 10 (3 marks)

Neil invests $3045 in an account that earns 3% p.a. compounded quarterly for 4 years.
David wants to invest his money in another account that earns simple interest at 3% p.a.
for 4 years.

If David is to earn the same amount of interest as Neil, how much must David invest in
his chosen account?

3 marks

Question 11 (4 marks)

Li buys 400 shares in FGH Industries at $3.80 per share. He pays a 1.8% brokerage fee
on the purchase. The shares have a dividend yield of 2.9%. A few months later, he sells the
shares for $4.20 per share and pays 1.2% brokerage on the sale of the shares.

a   What was the total cost of purchasing the shares?   1 mark

b   What is the value of the dividend Li receives?   1 mark

c   How much profit does Li make from selling his shares?   2 marks

Question 12 (2 marks)

Lincoln works for an insurance company and in 1992 he earned an annual salary of $85 000.

If his salary increased by 3% p.a. each year, how much more did he earn in 2020?   2 marks

Question 13 (3 marks)

Perla is planning to retire in 20 years. She wants to save enough money in a retirement fund to allow her to live on $3000 a month for 15 years after retirement.

a How much does Perla need to save for her retirement fund?   1 mark

b To reach her goal, Perla wants to contribute a certain amount to an annuity every 6 months that pays her 4% per half-year.

Use the future value table to find how much she will need to contribute every 6 months   2 marks
to give her enough money for retirement.

Future value of an annuity of $1 per period at the end of n periods

| Period | \multicolumn{10}{c}{Interest rate per period} |
	1%	2%	3%	4%	5%	6%	7%	8%	9%	10%
30	34.7849	40.5681	47.5754	56.0849	66.4388	79.0582	94.4608	113.2832	136.3075	164.4940
31	36.1327	42.3794	50.0027	59.3283	70.7608	84.8017	102.0730	123.3459	149.5752	181.9434
32	37.4941	44.2270	52.5028	62.7015	75.2988	90.8898	110.2182	134.2135	164.0370	201.1378
33	38.8690	46.1116	55.0778	66.2095	80.0638	97.3432	118.9334	145.9506	179.8003	222.2515
34	40.2577	48.0338	57.7302	69.8579	85.0670	104.1838	128.2588	158.6267	196.9823	245.4767
35	41.6603	49.9945	60.4621	73.6522	90.3203	111.4348	138.2369	172.3168	215.7108	271.0244
36	43.0769	51.9944	63.2759	77.5983	95.8363	119.1209	148.9135	187.1021	236.1247	299.1268
37	44.5076	54.0343	66.1742	81.7022	101.6281	127.2681	160.3374	203.0703	258.3759	330.0395
38	45.9527	56.1149	69.1594	85.9703	107.7095	135.9042	172.5610	220.3159	282.6298	364.0434
39	47.4123	58.2372	72.2342	90.4091	114.0950	145.0585	185.6403	238.9412	309.0665	401.4478
40	48.8864	60.4020	75.4013	95.0255	120.7998	154.7620	199.6351	259.0565	337.8824	442.5926
41	50.3752	62.6100	78.6633	99.8265	127.8398	165.0477	214.6096	280.7810	369.2919	487.8518
42	51.8790	64.8622	82.0232	104.8196	135.2318	175.9505	230.6322	304.2435	403.5281	537.6370
43	53.3978	67.1595	85.4839	110.0124	142.9933	187.5076	247.7765	329.5830	440.8457	592.4007
44	54.9318	69.5027	89.0484	115.4129	151.1430	199.7580	266.1209	356.9496	481.5218	652.6408
45	56.4811	71.8927	92.7199	121.0294	159.7002	212.7435	285.7493	386.5056	525.8587	718.9048

Questions 7–13 are worth 18 marks in total (Section II halfway point)

Question 14 (3 marks)

Alexander receives the following credit card statement.

Tick Bank	**STATEMENT OF ACCOUNT**
Where everything checks out	

Account Number: 111-234-567-890
Statement Date: 22/10/2020
Period Covered: 20/08/2020 to 20/10/2020   Page 1 of 1

Alexander Cruz

33 Espanol Drive
Elderslie NSW
2570

Camden Branch

Opening Balance:	$0.00
Total Credit Amount:	$870.00
Annual Percentage Rate:	16.8% p.a.
Closing Balance:	$870.00
Account Type:	Credit Account
Number of Transactions:	1

Transactions

Date	Description	Credit	Debit	Balance
05/09/2020	Eyephone10 – Mobiles'R'Us – Sydney	$870.00		$870.00

---End of transactions ---

Compound interest is charged on amounts from and including the date of payment up to and including the date of payment.

a Alexander pays his statement in full on 22 October 2020. How many days of interest   1 mark
   is he charged?

b What was the total amount owing on 22 October 2020?   1 mark

c How much interest did Alexander pay?   1 mark

Question 15 (3 marks)

Dan invests $3200 into an account in which interest compounds annually.
After 14 years, the value of his investment has grown to $6953.44.

What was the annual rate of interest of the investment, correct to one decimal place?   3 marks

_____

_____

_____

_____

_____

_____

_____

_____

_____

Question 16 (4 marks)

Lauren borrows $550 000 to buy her first house. The interest rate on the account is 3.5% p.a. compounded monthly. She is to repay $2754 each month for 25 years.

The balance at the end of 2 months is shown below.

Month	Principal at start of month	Interest	Repayment	Balance at end of month
1	$550 000	$1604.17	$2754.00	$548 850.17
2	$548 850.17	A	$2754.00	B

a   Calculate the values of A and B.   2 marks

_____

_____

_____

_____

_____

b   How much interest will Lauren pay in total over 25 years?   2 marks

_____

_____

_____

_____

_____

Question 17 (3 marks)

A trailer was bought for $10 000 and depreciated using the declining-balance method. The graph below shows the value of the trailer over a number of years.

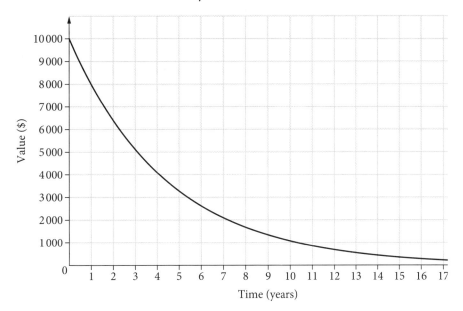

a Find the rate of depreciation per year as a percentage.   1 mark

b Calculate the value of the trailer after 18 years.   1 mark

c Explain why the value of the trailer will never be $0 according to this model.   1 mark

Question 18 (3 marks)

Harry opens an annuity account and contributes $2300 at the end of every 6 months. The interest rate on the account is 3% p.a. After $5\frac{1}{2}$ years, the interest rate increases to 4% p.a. He makes 2 more contributions at this new rate.

How much money will be in his account at the end of $6\frac{1}{2}$ years?   3 marks

Future value of an annuity of $1 per period at the end of *n* periods

	Interest rate per period									
Period	0.25%	0.50%	0.75%	1.00%	1.50%	2.00%	2.50%	3.00%	4.00%	5.00%
1	1.00000	1.0000	1.0000	1.0000	1.0000	1.0000	1.0000	1.0000	1.0000	1.0000
2	2.0025	2.0050	2.0075	2.0100	2.0150	2.0200	2.0250	2.0300	2.0400	2.0500
3	3.0075	3.0150	3.0226	3.0301	3.0452	3.0604	3.0756	3.0909	3.1216	3.1525
4	4.0150	4.0301	4.0452	4.0604	4.0909	4.1216	4.1525	4.1836	4.2465	4.3101
5	5.0251	5.0503	5.0756	5.1010	5.1523	5.2040	5.2563	5.3091	5.4163	5.5256
6	6.0376	6.0755	6.1136	6.1520	6.2296	6.3081	6.3877	6.4684	6.6330	6.8019
7	7.0527	7.1059	7.1595	7.2135	7.3230	7.4343	7.5474	7.6625	7.8983	8.1420
8	8.0704	8.1414	8.2132	8.2857	8.4328	8.5830	8.7361	8.8923	9.2142	9.5491
9	9.0905	9.1821	9.2748	9.3685	9.5593	9.7546	9.9545	10.1591	10.5828	11.0266
10	10.1133	10.2280	10.3443	10.4622	10.7027	10.9497	11.2034	11.4639	12.0061	12.5779
11	11.1385	11.2792	11.4219	11.5668	11.8633	12.1687	12.4835	12.8078	13.4864	14.2068
12	12.1664	12.3356	12.5076	12.6825	13.0412	13.4121	13.7956	14.1920	15.0258	15.9171

END OF PAPER

WORKED SOLUTIONS

Section I   (1 mark each)

Question 1

C $FV = PV(1 + r)^n$

$\qquad = 17\,000(1 + 0.0245)^{18}$

$\qquad = \$26\,282.35$

Straightforward question. Pay close attention to the compounding period.

Question 2

B $\qquad FV = PV(1 + r)^n$

$\qquad 188.28 = 100(1 + r)^{26}$

$\qquad (1 + r)^{26} = \dfrac{188.28}{100}$

$\qquad 1 + r = \sqrt[26]{1.8828}$

$\qquad r = \sqrt[26]{1.8828} - 1$

$\qquad\qquad = 0.0246\ldots$

$\qquad\qquad \approx 2.46\%$

Practise rearranging the formula to find either PV or r. This is an important skill to learn.

Question 3

C $\dfrac{101.591}{3220} = \$31.55$ per share

Dividend yield $= \dfrac{\text{dividend}}{\text{market price}} \times 100$

$\qquad\qquad = \dfrac{2.80}{31.55} \times 100$

$\qquad\qquad = 8.9\%$

Even though this is a straightforward question, make sure you know how to find the dividend yield because it is not on the reference sheet.

Question 4

A $S = V_0(1 - r)^n$

$\qquad = 22\,000(1 - 0.122)^3$

$\qquad = \$14\,890.40$

Difference $= 22\,000 - 14\,890.40$

$\qquad\qquad = \$7109.60$

Therefore, the motorbike loses $7109.60 in value over 3 years.

Make sure to read what the question is asking. Finding the salvage value is straightforward; however, you are asked for the difference between the original price and the salvage value.

Question 5

B Total amount $= 380\left(1 + \dfrac{0.16}{365}\right)^{20}$

$\qquad\qquad\qquad = \$383.35$

Interest $= \$383.35 - \$380.00$

$\qquad\quad = \$3.35$

Credit card questions always have small tricks in them. Make sure you convert the annual rate to a daily rate and know how to calculate the number of days of interest charged.

Question 6

D Interest $= \$200\,000 \times \dfrac{0.08}{12}$

$\qquad\qquad = \$1333.33$

Balance $= \$200\,000 + \$1333.33 - \$2030$

$\qquad\qquad = \$199\,303.33$

This common question requires you to know that interest is added before the repayment is taken out to obtain the balance at the beginning of the next period.

Section II ($\checkmark$ = 1 mark)

Question 7 (2 marks)

a Total dividend = 800×0.36
$$= \$288 \ \checkmark$$

b Dividend yield = $\dfrac{0.36}{3.80} \times 100$
$$= 9.5\% \ \checkmark$$

> This question is an easy introductory question where the answer to part **a** is used to answer part **b**. It is important to understand the definition of a dividend and a dividend yield.

Question 8 (2 marks)

Let V_0 be the purchase price of Ali's TV.
$$S = V_0(1 - r)^n$$
$$4686.63 = V_0(1 - 0.18)^3 \ \checkmark$$
$$V_0 = \frac{4686.63}{(0.82)^3}$$
$$= \$8500 \ \checkmark$$

> Finding V_0 (or the initial value of PV) is a common question. Make sure you know how to manipulate this equation algebraically.

Question 9 (2 marks)

Number of days charged interest = 52 $\checkmark$

$$A = P(1 + r)^n = 985(1 + 0.000\,443\,8)^{52}$$
$$\approx 1007.99$$

Interest = $\$1007.99 - \$985$
$$= \$22.99 \ \checkmark$$

> Look for little twists at the end. Notice that the question does not ask for the final balance but for the amount of interest charged.

Question 10 (3 marks)

Neil: $FV = PV(1 + r)^n$
$$= 3045(1 + 0.0075)^{16}$$
$$= \$3431.69 \ \checkmark$$

Interest = $\$3431.69 - \$3045$
$$= \$386.69$$

David: $I = Prn$
$$386.69 = P(0.03)(4) \ \checkmark$$
$$= 0.12P$$
$$P = \frac{386.69}{0.12}$$
$$\approx \$3222.42 \ \checkmark$$

Therefore, David must invest $3222.42 to receive the same amount of interest as Neil.

> This question involves the transfer of a result into another formula as well as algebraic manipulation. This makes it harder than a regular FV–PV question.

Question 11 (4 marks)

a Total cost to purchase
$$= (400 \times 3.80) + (400 \times 3.80 \times 0.018)$$
$$= \$1547.36 \ \checkmark$$

b Dividend = $400 \times 3.80 \times 0.029$
$$= \$44.08 \ \checkmark$$

c Profit = $(400 \times 4.20) + 44.08 - 1547.36 -$
$$(400 \times 4.20 \times 0.012) \ \checkmark$$
$$= \$156.56 \ \checkmark$$

> This question is not difficult but it can be very messy. Make sure that you use a logical and methodical approach when setting out your working.

Question 12 (2 marks)
$$FV = PV(1 + r)^n$$
$$= 85\,000(1 + 0.03)^{28}$$
$$= \$194\,473.85 \ \checkmark$$

Difference = $\$194\,473.85 - \$85\,000$
$$= \$109\,473.85 \ \checkmark$$

> This is a standard appreciation question. Again, look out for small twists such as the words 'how much more …'

Question 13 (3 marks)

a $3000 \times 12 \times 15 = \$540\,000 \ \checkmark$

b 20 years = 40 half-years. The future value interest factor for 40 periods at 4% per period
$$= 95.0255$$

Future value = contribution (C) × future value interest factor

$\$540\,000 = C \times 95.0255 \ \checkmark$
$$C = \frac{540\,000}{95.0255}$$
$$= \$5682.69 \text{ every 6 months} \ \checkmark$$

> The first part looks complicated but is an easy multiplication. In the second part, it is important to know that future value is contribution multiplied by interest factor. Learn how to use annuity tables correctly.

Question 14 (3 marks)

a Total days from 05/09/2020 to 22/10/2020
$$= 30 - 5 + 1 + 22$$
$$= 48 \text{ days } \checkmark$$

b $A = 870\left(1 + \dfrac{0.168}{365}\right)^{48}$
$$= \$889.43 \checkmark$$

c Interest $= \$889.43 - \$870$
$$= \$19.43 \checkmark$$

Credit card statements always contain a lot of information that is important to the question. Pay close attention to the dates, the number of days the balance is outstanding and the percentage rate either given as an annual rate or a daily rate. These questions are generally not answered well.

Question 15 (3 marks)
$$FV = PV(1 + r)^n$$
$$6953.44 = 3200(1 + r)^{14}$$
$$(1 + r)^{14} = \frac{6953.44}{3200} \checkmark$$
$$1 + r = \sqrt[14]{\frac{6953.44}{3200}} \checkmark$$
$$r = \sqrt[14]{\frac{6953.44}{3200}} - 1$$
$$= 0.056999\ldots$$
$$\approx 5.7\% \checkmark$$

This is a complex problem. Make sure that you know how to manipulate any equation in Mathematics Standard 2 to find any variable in an equation.

Question 16 (4 marks)

a $A = 548\,850.17 \times \dfrac{0.035}{12}$
$$= \$1600.81 \checkmark$$
$$B = 548\,850.17 + 1600.81 - 2754$$
$$= \$547\,696.98 \checkmark$$

b Total paid $= 2754 \times 12 \times 25$
$$= \$826\,200 \checkmark$$
Total interest $= \$826\,200 - \$550\,000$
$$= \$276\,200 \checkmark$$

Be familiar with tables showing reducing balance loans and how they work. Part **b** is a common question and is an easy way to work out the interest charged.

Question 17 (3 marks)

a Value at start $= \$10\,000$
Value in year 1 $= \$8000$
Depreciation $= \$10\,000 - \$8000$
$$= \$2000$$
Depreciation rate $= \dfrac{2000}{10\,000} \times 100 = 20\% \checkmark$

b $S = V_0(1 - r)^n = 10\,000(1 - 0.20)^{18}$
$$\approx \$180.14 \checkmark$$

c Since the balance is being decreased by a percentage, it will become smaller every year but cannot ever reach zero. $\checkmark$

Using graphs to interpret information is an important skill. Always look for points that are clearly on grid lines. This question is harder than a basic depreciation question because you are required to use the graph.

Question 18 (3 marks)

$r = 3\% \div 2 = 1.5\%$ every 6 months,
$n = 5.5$ years $= 11$ half-years

So future value interest factor $= 11.8633 \checkmark$

Future value
$= $ contribution $\times$ future value interest factor
$= 2300 \times 11.8633$
$= \$27\,285.59 \checkmark$

At the new rate of $4\% \div 2 = 2\%$ per half-year:

Balance (period 12) $= 27\,285.59 \times 1.02 + 2300$
$$= \$30\,131.30$$

Balance (period 13) $= 30\,131.30 \times 1.02 + 2300$
$$= \$33\,033.93 \checkmark$$

Therefore, Harry will have $\$33\,033.93$ in his account at the end of $6\frac{1}{2}$ years.

This is a complex financial mathematical problem. Annuities are often difficult to understand. Make sure you develop a good knowledge of the idea of an annuity. Remember that an annuity involves regular contributions. Know how to apply this to a future value table.

HSC exam topic grid (2011–2020)

This grid shows the coverage of this topic in past HSC exams by question number. The past exam papers can be downloaded from the NESA website (www.educationstandards.nsw.edu.au) by selecting 'Year 11 – Year 12', 'HSC exam papers'. NESA marking feedback and guidelines can also be found there.

Before 2019, 'Mathematics Standard 2' was called 'Mathematics General 2' and, before 2014, 'General Mathematics'. For these exams, select 'Year 11 – Year 12', 'Resources archive', 'HSC exam papers archive'.

	Investments and shares	Depreciation	Reducing balance loans	Annuities
2011	23(c)	28(b)	10*, 22	
2012	9	16, 26(b)	24, 26(c)*	
2013	9, 26(e), 28(d)	28(e)		
2014	30(a)	9		21
2015	17, 26(d)	10	29(a)*, 29(b)	30(c)
2016	8		17*, 27(d)	28(d)
2017	10, 26(e)	11	28(c)	27(c)
2018	19	26(h)	28(d)*, 29(e)	26(c)
2019 new course	3, 9, 13, 21	37	27*	42
2020	4, 21, 29	11	22*	14, 34, 37

* Credit cards.

CHAPTER 5
TOPIC EXAM

5

Bivariate data and the normal distribution

MS-S5 Bivariate data analysis

MS-S6 The normal distribution

- A reference sheet is provided on page 200 at the back of this book
- For questions in Section II, show relevant mathematical reasoning and/or calculations

Reading time: 4 minutes
Working time: 1 hour
Total marks: 40

Section I – 6 questions, 6 marks
- Attempt Questions 1–6
- Allow about 10 minutes for this section

Section II – 12 questions, 34 marks
- Attempt Questions 7–18
- Allow about 50 minutes for this section

Section I

• Attempt Questions 1–6 • Allow about 10 minutes for this section	**6 marks**

Question 1

Which correlation coefficient value best represents the relationship shown in the scatterplot below?

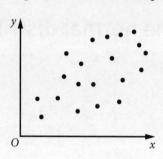

A −0.88

B −0.15

C 0.15

D 0.88

Question 2

Which graph best represents a data set with a small standard deviation and a median that is greater than its mean?

A

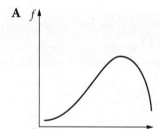

B

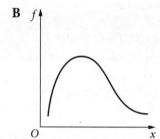

C

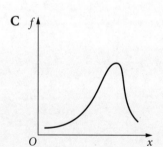

D

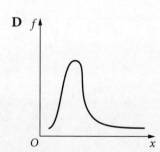

Question 3

Which graph describes a correlation coefficient closest to 0?

A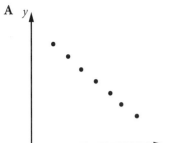

B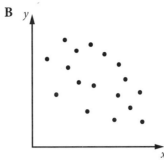

C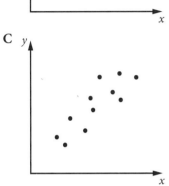

D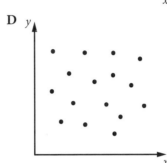

Question 4

The scores in a recent Science test are normally distributed with a mean of 62 and a standard deviation of 5.

What is the z-score for a mark of 77?

A −3

B −2

C 2

D 3

Question 5

Which terms best describe the relationship shown in the scatterplot below?

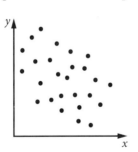

A Weak, negative

B Weak, positive

C Strong, negative

D Strong, positive

Question 6

The test results for a group of university students are normally distributed with a mean of 65 and a standard deviation of 8.

What percentage of students achieved a mark greater than 89?

A 0.15%

B 0.3%

C 34%

D 99.7%

Section II

• Attempt Questions 7–18 • Allow about 50 minutes for this section • Answer the questions in the spaces provided. These spaces provide guidance for the expected length of response. • Your responses should include relevant mathematical reasoning and/or calculations.	**34 marks**

Question 7 (2 marks)

The lengths of insects of a particular type are found to be normally distributed with a standard deviation of 3.2 mm. An insect that is 21 mm long has a z-score of 1.8.

What is the mean of the distribution of insect lengths?   2 marks

Question 8 (4 marks)

Duncan's weight was recorded every 5 years of his life for 45 years.

Age	20	25	30	35	40	45	50	55	60	65
Weight (kg)	67	63	65	69	78	81	77	82	85	87

a What is the correlation coefficient of the data, correct to four decimal places?   1 mark

b Describe the relationship between the two variables.   1 mark

c What do you expect to happen when Duncan turns 70?   1 mark

d Is this a causal relationship? Give a reason for your answer.   1 mark

Question 9 (2 marks)

Mehrnoosh scores 52 on her Biology test and 68 on her Physics test. The class results of both tests are normally distributed, with the means and standard deviations shown.

	Mean	Standard deviation
Biology	45	5
Physics	59	7.5

Mehrnoosh claims that she performed better in Biology. Is she correct? Use appropriate calculations to justify your answer.   2 marks

Question 10 (3 marks)

The heights and arm spans of 7 swimmers are shown in the table.

Height (h cm)	185	188	189	192	195	194	193
Arm span (s cm)	187	189	192	193	197	195	193

a Which of the variables is the independent variable?   1 mark

b Calculate Pearson's correlation coefficient for the data set, correct to four decimal places.   1 mark

c Describe the relationship between the two variables.   1 mark

Question 11 (2 marks)

Kyren achieves a z-score of 2.3 in his English test and a z-score of 2.1 in his Science test. The class results of both tests are normally distributed, with their means and standard deviations shown in the table.

	Mean	Standard deviation
English	65.5	5
Science	59.6	4

What is the difference in Kyren's actual test results for the two subjects?   2 marks

_____

_____

_____

_____

Question 12 (3 marks)

The scatterplot below shows the ages and heights of 6 girls.

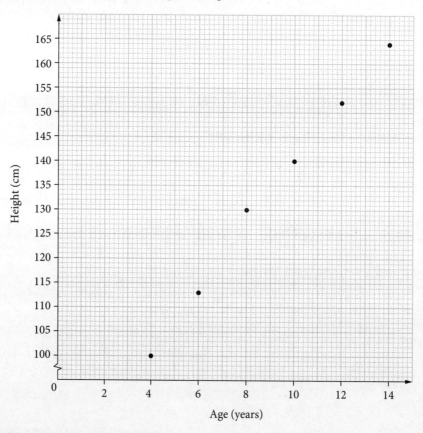

a   Which variable is the dependent variable?   1 mark

_____

b   Calculate Pearson's correlation coefficient for the data, correct to four decimal places.   1 mark

_____

c   Describe the relationship between the two variables.   1 mark

_____

_____

Question 13 (2 marks)

The results of 200 students who took a Maths test are normally distributed. The mean is 81 and the standard deviation is 3.

If one student is chosen at random from the 200 students, what is the probability of choosing   2 marks
a student who attained a score between 84 and 87?

Questions 7–13 are worth 18 marks in total (Section II halfway point)

Question 14 (2 marks)

The numbers of paperclips in boxes produced by a factory are found to be normally distributed with a mean of 198 and a standard deviation of 2.

What percentage of boxes will have between 196 and 202 paperclips?   2 marks

TOPIC EXAM

Question 15 (3 marks)

Data for temperature, T, and rainfall, R, were measured and recorded in a particular city over 8 days.
The data was placed in a scatterplot and a line of best fit was drawn.

Temperature ($T°$C)	27	18	22	25	19	21	26	24
Rainfall (R mm)	110	45	80	105	60	90	75	95

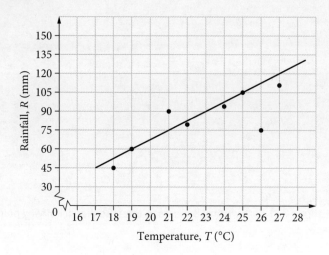

a Determine the equation of the line of best fit shown.   2 marks

_____

_____

_____

_____

_____

_____

b Use the equation to predict how much rainfall would occur if the temperature reached 23°C.   1 mark

_____

_____

_____

Question 16 (5 marks)

The waist and chest sizes of 10 males were measured. The results are shown in the table below.

Waist (*W* cm)	64	79	90	78	75	69	88	89	72	79
Chest (*C* cm)	83	94	102	42	91	82	99	103	74	93

a Describe the strength of the relationship of the two variables using Pearson's   2 marks
   correlation coefficient. Give your answer correct to four decimal places.

b Find the equation of the least-squares regression line. Round your answers to four   2 marks
   decimal places.

c Use the equation of the least-squares regression line to predict the waist size of a male with   1 mark
   a chest size of 100 cm, correct to two decimal places.

Question 17 (2 marks)

The final examination results of 600 students are normally distributed, with a mean of 63 and
a standard deviation of 6.

How many students scored less than 51?   2 marks

TOPIC EXAM

Question 18 (4 marks)

A set of scores is normally distributed. Out of 2000 scores, only 3 of the scores are greater than 28. The standard deviation of the data set is 2.

a   What is the mean of the set of scores?   2 marks

_____

_____

_____

_____

b   If a score is picked at random, what is the probability (as a simplified fraction) of   2 marks
choosing a score that is 1 standard deviation from the mean?

_____

_____

_____

_____

END OF PAPER

WORKED SOLUTIONS

Section I (1 mark each)

Question 1

C Points are spread further apart (weak correlation) but are still trending in a positive direction.

> Look for both direction and spread of the points in a scatterplot.

Question 2

C A smaller standard deviation means the data is bunched closer together and results in a narrower peak of the graph. When the median is greater than the mean, a negative skew will result.

> A negative skew drags the mean to the left, so it is lower than the median (and mode/peak). The opposite occurs for a positive skew.

Question 3

D Remember that a correlation coefficient will range from −1 to 1. A correlation coefficient of 0 means no correlation so the points are random and show no relationship between the variables.

> This question is straightforward and relies on your understanding of the correlation coefficient and how it relates to a scatterplot.

Question 4

D $z = \dfrac{x - \mu}{\sigma}$

$ = \dfrac{77 - 62}{5}$

$ = 3$

> Be familiar with the z-score formula, which is given on the reference sheet.

Question 5

A The points in the scatterplot are trending downward, meaning it is a negative relationship. The points are spread apart and not close together, therefore the relationship is weak.

> Describing the relationship shown in a scatterplot is an important skill in this topic.

Question 6

A

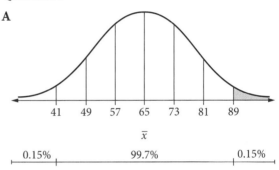

89 is 3 standard deviations above the mean. Shaded percentage is $(100\% - 99.7\%) \div 2 = 0.15\%$. Therefore, 0.15% of students achieved a mark greater than 89.

> Be familiar with the percentages in a normal distribution. The empirical rule is on the reference sheet but some students like to memorise the areas of the individual regions to make calculations easier in problems. The area from $z = 0$ to 1 is 34% ($\frac{1}{2}$ of 68%), then moving right the areas are 13.5%, 2.35% and 0.15%. It is the same going from 0 to −1 and so on.

Section II ($\checkmark$ = 1 mark)

Question 7 (2 marks)

$$z = \frac{x - \mu}{\sigma}$$

$$1.8 = \frac{21 - \mu}{3.2}$$

$$3.2 \times 1.8 = 21 - \mu \quad \checkmark$$

$$\mu = 21 - 5.76$$

$$= 15.24 \, \text{mm} \quad \checkmark$$

> Be prepared to find any variable in the z-score formula. This will require you to practise your algebraic skills.

Question 8 (4 marks)

a $r = 0.9385$ (using the statistics mode on the calculator) $\checkmark$

b The correlation coefficient indicates a strong, positive linear relationship. $\checkmark$

c As there is a strong, positive linear relationship, an increase in one variable should lead to an increase in the other variable. Therefore, his weight would be expected to continue to increase as his age increases to 70. $\checkmark$

d It is not a causal relationship as age does not necessarily cause an increase in weight. $\checkmark$

> It is important to know how to use your calculator in statistics mode to find the correlation coefficient or least-squares regression line.

Question 9 (2 marks)

$$z = \frac{x - \mu}{\sigma}$$

Biology:

$$z = \frac{52 - 45}{5}$$

$$= 1.4 \quad \checkmark$$

Physics:

$$z = \frac{68 - 59}{7.5}$$

$$= 1.2$$

Therefore, Mehrnoosh is correct because her z-score for Biology is higher than her z-score for Physics. $\checkmark$

> Comparing z-scores is a common question. Remember, the higher the z-score, the better the result.

Question 10 (3 marks)

a Height is the independent variable. $\checkmark$

> The independent variable is the variable on the horizontal axis of a graph and is usually in the top row of a table.

b $r = 0.9634$ $\checkmark$

c The relationship between the variables is a strong, positive linear relationship. $\checkmark$

Question 11 (2 marks)

$$z = \frac{x - \mu}{\sigma}$$

English:

$$2.3 = \frac{x - 65.5}{5}$$

$$11.5 = x - 65.5$$

$$x = 77 \quad \checkmark$$

Science:

$$2.1 = \frac{x - 59.6}{4}$$

$$8.4 = x - 59.6$$

$$x = 68$$

Therefore, the difference between the 2 marks is $77 - 68$, which is 9. $\checkmark$

> This question relies on your ability to use the z-score formula to find variables other than z. Always make sure that you answer all parts in the question.

Question 12 (3 marks)

a The dependent variable is height, because it 'depends on' age. $\checkmark$

b $r \approx 0.9975$ $\checkmark$

c The relationship between the 2 variables is a strong, positive linear relationship. $\checkmark$

> The dependent variable is on the vertical axis. Make sure you know how to enter data values in your calculator and then find the correlation coefficient.

Question 13 (2 marks)

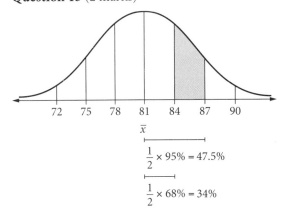

$\frac{1}{2} \times 95\% = 47.5\%$

$\frac{1}{2} \times 68\% = 34\%$

Shaded area = 47.5 − 34

= 13.5% ✓

13.5% of 200 = 27

Therefore, the probability of choosing a student

who scored between 84 and 87 is $\frac{27}{200}$. ✓

Understanding the percentages in a normal distribution is an important skill. The harder questions usually involve subtractions.

Question 14 (2 marks)

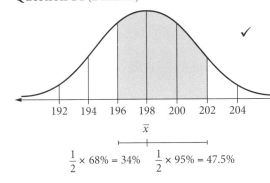

✓

$\frac{1}{2} \times 68\% = 34\%$ $\frac{1}{2} \times 95\% = 47.5\%$

Shaded area = 34 + 47.5

= 81.5%

Therefore, the percentage of boxes that contain between 196 and 202 paperclips is 81.5%. ✓

It is always helpful to draw a graph of the normal distribution to visualise the required percentages between values. Some students like to memorise the areas of the individual regions to make calculations easier. The area from $z = 0$ to 1 is 34% ($\frac{1}{2}$ of 68%), then moving right the areas are 13.5%, 2.35% and 0.15%. So the shaded area is 34% + 34% + 13.5% = 81.5%.

Question 15 (3 marks)

a

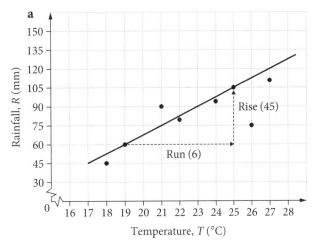

Choosing $(19, 60)$ and $(25, 105)$ on the line:

$m = \dfrac{\text{rise}}{\text{run}}$

$= \dfrac{45}{6}$

$= 7.5$ ✓

$y = mx + c$

$R = 7.5T + c$

When $T = 25$, $R = 105$

So $105 = 7.5(25) + c$

$105 = 187.5 + c$

$c = -82.5$

So $R = 7.5T - 82.5$ ✓

b When $T = 23$:

$R = 7.5(23) - 82.5$

$= 90 \,\text{mm}$ ✓

When finding the equation of a straight line, both the gradient and the value of c must be found. Find the gradient by picking 2 clear-cut points on the line and find c by substituting a point on the line into your equation with your found gradient.

Question 16 (5 marks)

a $r = 0.4614$ ✓

This shows a moderate, positive linear relationship ✓
between waist size and chest size.

b Using STAT mode on a calculator:

$a = 11.4137$ and $b = 0.9564$ ✓

So $y = bx + a$

$C = 0.9564W + 11.4137$ ✓

c If $C = 100$:

$100 = 0.9564W + 11.4137$

$88.5863 = 0.9456W$

$W = \dfrac{88.5863}{0.9456}$

$= 92.63 \text{ cm}$ ✓

> It is important to be familiar with finding both b and a from your calculator to find the gradient and y-intercept, respectively, of the least-squares regression line. This question requires accurate and efficient calculator skills.

Question 17 (2 marks)

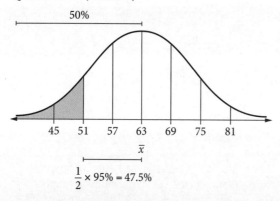

$\dfrac{1}{2} \times 95\% = 47.5\%$

Shaded area $= 50 - 47.5$

$= 2.5\%$ ✓

$2.5\% \times 600 = 15$ students ✓

> Remember to look carefully at what the question is asking rather than giving the correct percentage as your final answer.

Question 18 (4 marks)

a $\dfrac{3}{2000} \times 100 = 0.15\%$ ✓

0.15% is more than 3 standard deviations above the mean: $(100\% - 99.7\%) \div 2 = 0.15\%$

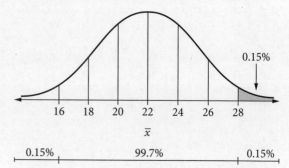

28 is 3 standard deviations above the mean and 1 standard deviation = 2, therefore the mean is $28 - 3 \times 2 = 22$. ✓

> This question is difficult as it relies on your ability to work backwards using your knowledge of the normal distribution. Only 3 scores are higher than 28, so those 3 scores must be the highest scores in the data set. The percentage of scores higher than 28 gives the clue about the rest of the distribution.

b One standard deviation either side of the mean represents 68% of the data.

So probability $= \dfrac{68}{100}$ ✓

$= \dfrac{17}{25}$ ✓

HSC exam topic grid (2011–2020)

This grid shows the coverage of this topic in past HSC exams by question number. The past exams can be downloaded from the NESA website (www.educationstandards.nsw.edu.au) by selecting 'Year 11 – Year 12', 'HSC exam papers'. NESA marking feedback and guidelines can also be found there.

Before 2019, 'Mathematics Standard 2' was called 'Mathematics General 2' and, before 2014, 'General Mathematics'. For these exams, select 'Year 11 – Year 12', 'Resources archive', 'HSC exam papers archive'.

	Scatterplots and correlation	Linear regression	The normal distribution	z-scores
2011	8			27(c)
2012	11, 29(a)	19 (replace 'median regression' with 'line of best fit')	29(b)	
2013	2	28(b)	20	29(b)
2014	30(b)(i)	30(b) except (v)	24	
2015	28(e)	28(e)	20	28(b)
2016	3, 29(d)(i)	29(d)(ii) using table of values	13	
2017	12		29(d)	13
2018		29(d) (use table of values for (i))	23	27(e)
2019 new course	23(a)–(b)	23(c)	15	38
2020	12	36	35	8, 35

CHAPTER 6
TOPIC EXAM

Networks

• A reference sheet is provided on page 200 at the back of this book • For questions in Section II, show relevant mathematical reasoning and/or calculations	**Reading time: 4 minutes** **Working time: 1 hour** **Total marks: 40**

Section I – 6 questions, 6 marks
• Attempt Questions 1–6
• Allow about 10 minutes for this section

Section II – 11 questions, 34 marks
• Attempt Questions 7–17
• Allow about 50 minutes for this section

Section I

• Attempt Questions 1–6	**6 marks**
• Allow about 10 minutes for this section	

Question 1

What is the sum of the degrees of all vertices in the network below?

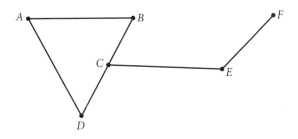

A 10 **B** 12 **C** 14 **D** 16

Question 2

Water pipes connect 8 locations. The flow network below shows the capacities of the pipes.

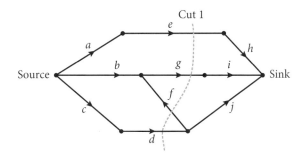

What is the capacity of cut 1?

A $e + d$ **B** $e + g + f$ **C** $e + g + d$ **D** $e + g - f + d$

Question 3

What is the weight of the minimum spanning tree in the network below?

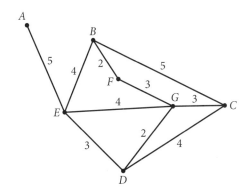

A 17 **B** 18 **C** 19 **D** 22

Question 4

Which of the following statements is true regarding networks A and B?

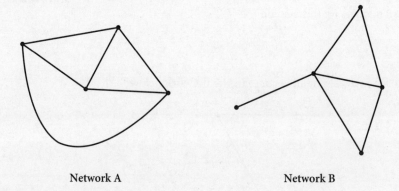

Network A   Network B

A The sum of the degrees of Network B is greater than that of Network A.

B Network B has 1 more vertex with an odd degree than Network A.

C Network B has fewer vertices than Network A.

D The sum of the degrees of Network B is the same as that of Network A.

Question 5

In the network below, points *A* and *B* are connected through a series of paths. The values on each path represent distances, in metres.

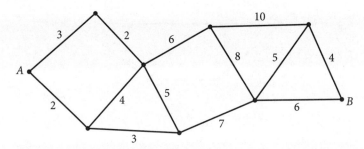

What is the shortest distance between points *A* and *B*?

A 17 m   **B** 18 m

C 25 m   **D** 26 m

Question 6

Find the critical path of the network below.

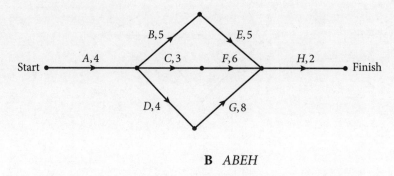

A *ADGH*   **B** *ABEH*

C *ACFH*   **D** *ACH*

9780170459211

Section II

> - Attempt Questions 7–17   **34 marks**
> - Allow about 50 minutes for this section
> - Answer the questions in the spaces provided. These spaces provide guidance for the expected length of response.
> - Your responses should include relevant mathematical reasoning and/or calculations.

Question 7 (2 marks)

A rectangular field is separated into 6 different areas for growing different crops.

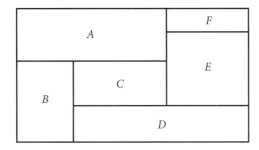

Complete the network of the field shown below, indicating the connections between fields that are in contact with each other.   2 marks

Question 8 (3 marks)

The network below shows distances, in metres, between 9 locations.

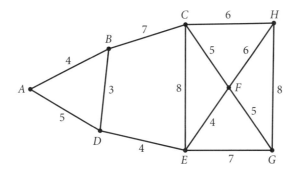

a   What is the shortest distance between A and H?   1 mark

b Does the shortest path lie on the network's minimum spanning tree?   2 marks
Show your answer using the diagram below.

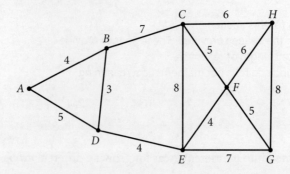

_____

_____

Question 9 (4 marks)

Towns D, E, F and G are connected by roads, and the distances, in kilometres, are shown in the grid.

	D	E	F	G
D	–	14	22	16
E	14	–	41	12
F	22	41	–	27
G	16	12	27	–

a Draw a weighted network diagram that represents the information in the grid.   2 marks

b A trucking company is based in town F and wants to visit each town, returning to F along
the shortest path.

What route should the company take and what is the length of the route?   2 marks

_____

_____

_____

_____

Question 10 (3 marks)

The following network joins 5 vertices.

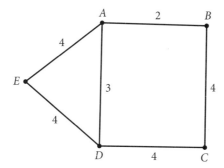

a How many minimum spanning trees are possible?   1 mark

_____

_____

_____

_____

_____

b Draw 1 minimum spanning tree and indicate its length.   2 marks

9780170459211

Question 11 (2 marks)

The following plan is for a small flat being built in a backyard.

Draw a network that models the house plan. Show areas as vertices, and show doorways and entrances as edges.   2 marks

Question 12 (3 marks)

A company wants to install internet cabling to connect 6 suburbs. The network below shows the length, in metres, of cabling required to connect each suburb.

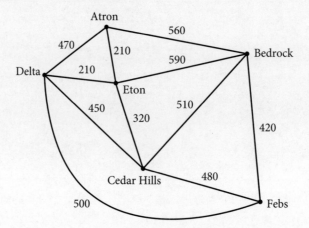

What is the minimum cost of connecting all 6 suburbs if cabling costs $50 per metre?   3 marks

Questions 7–12 are worth 17 marks in total (Section II halfway point)

Question 13 (3 marks)

The network diagram shows tasks that must be completed to finish a project. Each edge weight represents the time taken to complete the task, in days.

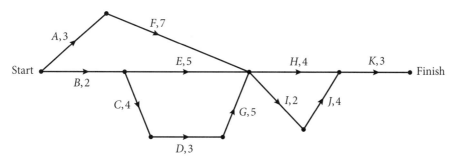

a What is the estimated start time (EST) of activity H?   1 mark

b What is the critical path for the project?   1 mark

c What is the float time of activity E?   1 mark

Question 14 (4 marks)

The table shows the activities of a project, their durations, in days, and their immediate predecessors.

Activity	Duration (days)	Immediate predecessor(s)
A	3	–
B	4	A
C	4	–
D	5	C
E	7	C
F	6	C
G	4	B, F
H	4	D
I	7	E, G, H

a Using the table, complete the network below. 2 marks

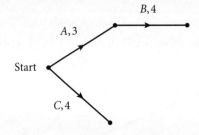

b What is the minimum completion time for the project? 1 mark

c What is the float time of activity F? 1 mark

Question 15 (3 marks)

The network below shows the flow of water through a series of pipes, from a source to the sink.
Each number refers to the capacity of a pipe, in kilolitres per hour.

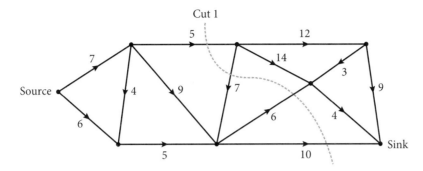

a What is the capacity of cut 1?   1 mark

_____

_____

b What is the maximum flow of the network, in kilolitres per hour?   2 marks

_____

_____

_____

TOPIC EXAM

Question 16 (4 marks)

Annette wants to plan a trip to visit 8 different cities over the Christmas break. She begins her trip from *A*. The costs to travel from one city to another are given as edge weights.

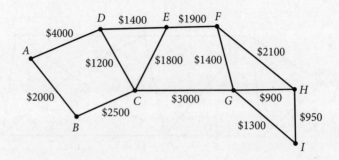

a   If Annette wants to visit all 8 cities, what is the least amount of money she will need to cover her travel expenses?   1 mark

_____

_____

_____

_____

b   What is the cheapest route from *A* to *I* that Annette can take, and how much will this cost?   2 marks

_____

_____

_____

_____

_____

c   Why is the cheapest route from *A* to *I* NOT the best path to take from a cost perspective if Annette still wants to visit the other 7 cities?   1 mark

_____

_____

_____

_____

Question 17 (3 marks)

A project has 9 activities, with durations measured in weeks. This is shown in the directed network below.

How many activities would need to be reduced by 2 weeks to reduce the minimum completion   3 marks
time of the project to 21 weeks?

_____

_____

_____

_____

_____

_____

_____

_____

_____

_____

_____

_____

_____

END OF PAPER

WORKED SOLUTIONS

Section I (1 mark each)

Question 1

B $A(2)$, $B(2)$, $C(3)$, $D(2)$, $E(2)$, $F(1)$.

Sum = 12.

> Straightforward question. Know the meanings of basic terminology, such as 'degree'.

Question 2

C When determining the capacity of a cut, only add edges, leaving the source. Ignore edges heading back to the source, such as edge f.

> Remember that an edge that points back to the source is not counted in the capacity of the cut.

Question 3

B The minimum spanning tree connects all vertices with the smallest edge weights possible without forming cycles.

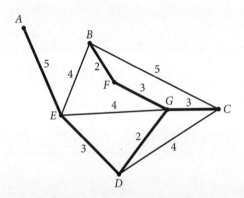

> Straightforward question. Make sure you are familiar with Kruskal's and Prim's methods for minimum spanning trees.

Question 4

D Sum of the degrees of Network A

= 3 + 3 + 3 + 3

= 12

Sum of the degrees of Network B

= 1 + 4 + 2 + 3 + 2

= 12

> This question tests your knowledge of various terms. Know the difference between 'odd degrees' and 'sum of the degrees'.

Question 5

B Shortest distance = 2 + 3 + 7 + 6 = 18.

> Finding the shortest path is a common exam question. With smaller networks like this one, it is usually quicker to find the shortest path by trial and error.

Question 6

A To find the critical path, find the EST and LST for each vertex. The critical path includes vertices where the EST and LST are the same.

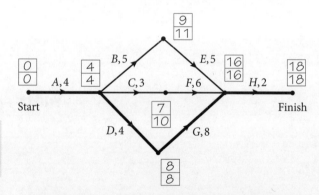

> Finding the critical path on a network is a skill that you should practise. This is a common question.

Section II ($\checkmark$ = 1 mark)

Question 7 (2 marks)

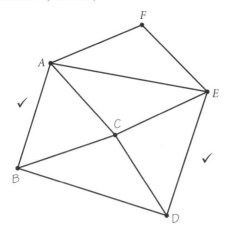

Always check to make sure that all connections have been represented.

Question 8 (3 marks)

a Shortest distance is 17 m, as indicated by the darkened path below. $\checkmark$

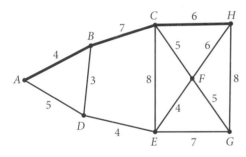

b Shaded path below indicates one of the minimum spanning trees. Therefore, the shortest path does not lie on the minimum spanning tree. $\checkmark$

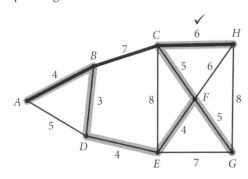

Be prepared to answer questions combining minimum spanning trees and shortest paths. These concepts themselves are regularly examined.

Question 9 (4 marks)

a

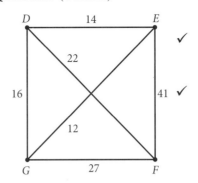

b The shortest path is *FDEGF*. $\checkmark$

The distance of the shortest path is 75 km. $\checkmark$

Drawing diagrams from tables is an important concept. This is a routine question.

Question 10 (3 marks)

a There are 4 possible minimum spanning trees. $\checkmark$

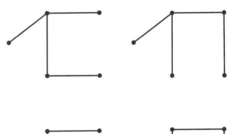

b Length of minimum spanning tree = 13 $\checkmark$

Example of a minimum spanning tree:

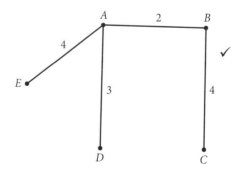

Be familiar with the 2 methods of finding minimum spanning trees. These questions are some of the easier networks problems.

Question 11 (2 marks)

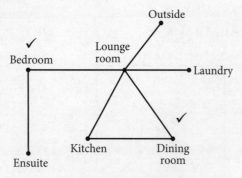

> Remember that when drawing a network from a house plan, a vertex for 'outside' needs to be included.

Question 12 (3 marks)

First, find the minimum spanning tree.

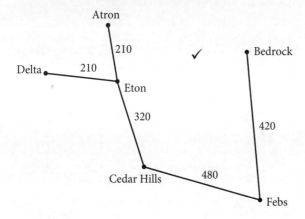

Total distance:

$210 + 210 + 320 + 480 + 420 = 1670\,\text{m} = 1640\,\text{m}$ ✓

Total cost $= 1640 \times 50 = \$82\,000$ ✓

> Make sure the question is answered completely to get full marks.

Question 13 (3 marks)

a EST of activity $H = 14$ days ✓

b Critical path:

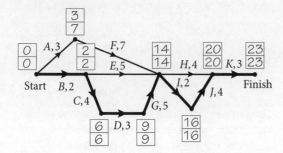

This means critical path is $BCDGIJK$. ✓

c Float time of activity $E =$ LST – EST – duration

$\qquad = 14 - 2 - 5$

$\qquad = 7$ days ✓

> Float time is an important concept. It is not just the difference between LST and EST, as part **c** shows.

Question 14 (4 marks)

a

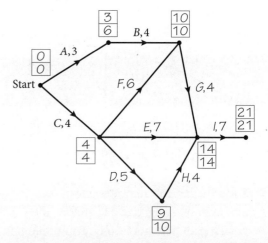

b The minimum completion time for the project is 21 days. ✓

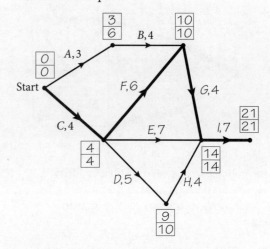

c The float time of activity F is 0 because it lies on the critical path. ✓

> Finding the critical path is an important concept to master.

Question 15 (3 marks)

a Capacity of cut 1 = 5 + 6 + 10
$$= 21 \, \text{kL/h} \; \checkmark$$

b Minimum cut = 7 + 5 = 12
Therefore, maximum flow = 12 kL/h ✓

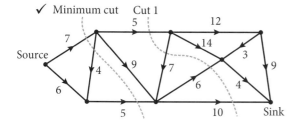

The minimum cut is equal to the maximum flow of the network.

Question 16 (4 marks)

a The minimum spanning tree will equal the least amount of money.

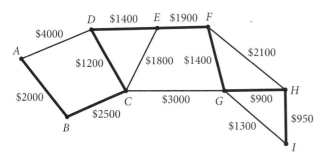

Total of all edges = 12 250, so least amount of money is $12 250. ✓

b Shortest path from A to I = ABCGI ✓

Cost = $2000 + $2500 + $3000 + $1300
$$= \$8800 \; \checkmark$$

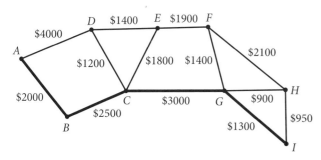

c The cheapest route is not the best path if all other cities are to be visited, because the path does not lie on the minimum spanning tree. ✓

Minimum path questions are straightforward and should be practised.

Question 17 (3 marks)

Minimum completion time = 29 weeks

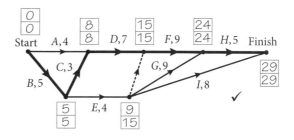

Look at the distance of all paths in this network, from start to finish:

$ADFH = 25$
$BCDFH = 29$ (critical path)
$BEFH = 23$
$BEGH = 17$
$BEI = 23$

To reduce the overall completion time, reduce activities on the critical path.

Be careful not to reduce activities to the point that a new critical path is formed. Look for activities that are common to all paths.

Reducing *H*:
$ADFH = 23$
$BCDFH = 27$ (critical path) ✓
$BEFH = 21$
$BEGH = 15$
$BEI = 23$

Reducing *F*:
$ADFH = 21$
$BCDFH = 25$ (critical path)
$BEFH = 19$
$BEGH = 15$
$BEI = 23$

Reducing *D*:
$ADFH = 19$
$BCDFH = 23$ (critical path)
$BEFH = 19$
$BEGH = 15$
$BEI = 23$

Reducing *C*:
$ADFH = 19$
$BCDFH = 21$ (critical path)
$BEFH = 19$
$BEGH = 15$
$BEI = 23$

This means *C* cannot be reduced because *BEI* replaces *BCDFH* as the new critical path.

Reducing B:

$ADFH = 19$

$BCDFH = 21$ (critical path)

$BEFH = 17$

$BEGH = 13$

$BEI = 21$

Even though a new critical path has been created, $BCDFH$ is still a critical path.

Therefore, reduce activities B, D, F and H by 2 weeks for each activity.

Therefore, 4 activities need to be changed to reduce the minimum completion time of the project to 21 weeks. ✓

> This process has been shown step-by-step to explain how to reduce activities. It is important to find all of the paths and to have a systematic and logical method of going through the process.

HSC exam topic grid (2019–2020)

This table shows the coverage of this topic in past HSC exams by question number. The past exams can be downloaded from the NESA website (www.educationstandards.nsw.edu.au) by selecting 'Year 11 – Year 12', 'HSC exam papers'. NESA marking feedback and guidelines can also be found there.

Networks was introduced to the Mathematics Standard 2 course in 2019.

	Terminology and diagrams	Minimum spanning trees	Shortest paths	Critical path analysis	Flow networks
2019 new course		30(a)	30(b)	26	40
2020	9	18		26	30

Mathematics Standard 2

PRACTICE MINI-HSC EXAM 1

General instructions	• Reading time: 4 minutes
	• Working time: 1 hour
	• A reference sheet is provided on page 200 at the back of this book
	• For questions in Section II, show relevant mathematical reasoning and/or calculations

Total marks: 40	**Section I – 6 questions, 6 marks**
	• Attempt Questions 1–6
	• Allow about 10 minutes for this section
	Section II – 13 questions, 34 marks
	• Attempt Questions 7–19
	• Allow about 50 minutes for this section

Section I

6 marks
Attempt Questions 1–6
Allow about 10 minutes for this section

Circle the correct answer.

Question 1

In the network below, how many vertices have an odd degree?

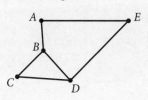

A 2

B 3

C 4

D 5

Question 2

Which of the following graphs could be described as having a correlation coefficient close to −0.8?

A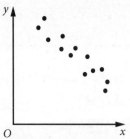

B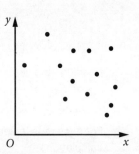

C

D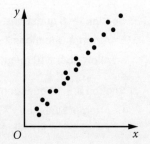

Question 3

A right-angled triangle has angles in the ratio $2:4:6$.

What is the size of the smallest angle?

A 15°

B 30°

C 45°

D 90°

Question 4

Which of the following best represents the graph of $y = 3^x$?

A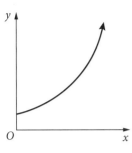

B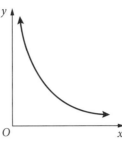

C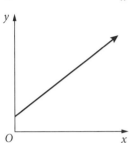

D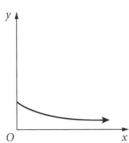

Question 5

Sasha knows that a basketball ring is 3.05 m high. He looks up at the front of the ring at an angle of elevation of 21°. His eyes are 160 cm from the ground.

How far is Sasha from standing directly underneath the front of the ring?

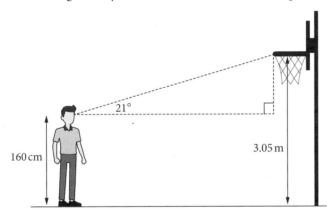

NOT TO SCALE

A 0.56 m

B 3.78 m

C 4.56 m

D 7.94 m

Question 6

Tom contributes $3000 per year for 8 years into an annuity that gives 2.5% p.a. interest.

How much interest does he accrue over 8 years? Use the table to calculate your answer.

Future value interest factors
(Future value of an annuity with a contribution of $1 at the end of each period)

Period	Interest rate per period									
	0.25%	0.50%	0.75%	1.00%	1.50%	2.00%	2.50%	3.00%	4.00%	5.00%
1	1.0000	1.0000	1.0000	1.0000	1.0000	1.0000	1.0000	1.0000	1.0000	1.0000
2	2.0025	2.0050	2.0075	2.0100	2.0150	2.0200	2.0250	2.0300	2.0400	2.0500
3	3.0075	3.0150	3.0226	3.0301	3.0452	3.0604	3.0756	3.0909	3.1216	3.1525
4	4.0150	4.0301	4.0452	4.0604	4.0909	4.1216	4.1525	4.1836	4.2465	4.3101
5	5.0251	5.0503	5.0756	5.1010	5.1523	5.2040	5.2563	5.3091	5.4163	5.5256
6	6.0376	6.0755	6.1136	6.1520	6.2296	6.3081	6.3877	6.4684	6.6330	6.8019
7	7.0527	7.1059	7.1595	7.2135	7.3230	7.4343	7.5474	7.6625	7.8983	8.1420
8	8.0704	8.1414	8.2132	8.2857	8.4328	8.5830	8.7361	8.8923	9.2142	9.5491
9	9.0905	9.1821	9.2748	9.3685	9.5593	9.7546	9.9545	10.1591	10.5828	11.0266
10	10.1133	10.2280	10.3443	10.4622	10.7027	10.9497	11.2034	11.4639	12.0061	12.5779
11	11.1385	11.2792	11.4219	11.5668	11.8633	12.1687	12.4835	12.8078	13.4864	14.2068
12	12.1664	12.3356	12.5076	12.6825	13.0412	13.4121	13.7956	14.1920	15.0258	15.9171

A $2208.30

B $3408.30

C $24 000.00

D $26 208.30

Section II

34 marks
Attempt Questions 7–19
Allow about 50 minutes for this section

- Answer the questions in the spaces provided. These spaces provide guidance for the expected length of response.
- Your responses should include relevant mathematical reasoning and/or calculations.

Question 7 (1 mark)

The diagram shows the positions of towns E, F and G. Town E is due north of town G and $\angle GEF = 28°$.

What is the bearing of town E from town F?   1 mark

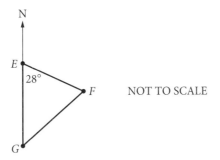

NOT TO SCALE

Question 8 (2 marks)

The normal distribution below has a mean of 120 and a standard deviation of 8.

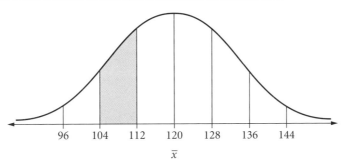

a What percentage of the data lies in the shaded region?   1 mark

b Write a possible z-score that lies in the shaded region.   1 mark

Question 9 (2 marks)

A computer program was used to graph Line A and Line B. One line has the equation $y = 3x - 8$. The other line has the equation $y = -3x - 2$.

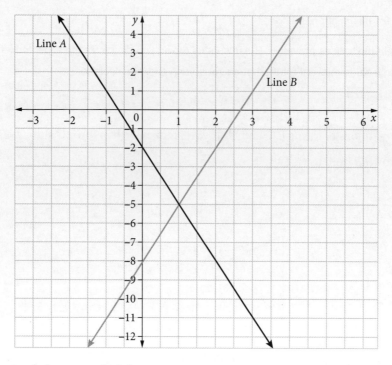

a   Specify which equation belongs to which line.   1 mark

_____

_____

_____

b   What is the solution to the pair of simultaneous equations?   1 mark

_____

_____

_____

Question 10 (2 marks)

The following table shows all of the activities that need to be completed for a project to finish on time.

Activity	Duration (hours)	Immediate predecessor(s)
A	5	–
B	7	–
C	4	B
D	5	A
E	4	C, D
F	7	B
G	2	E, F

What is the float time of activity *D*?   2 marks

Question 11 (2 marks)

Amalia completes a class test in two subjects. The class scores for each test are normally distributed. The table below shows Amalia's marks, as well as the mean and standard deviation for the class scores for each test.

Subject	Amalia's score	Mean	Standard deviation
Maths	71	57.8	6
English	79	64.3	7

Show by calculations which test Amalia did relatively better in.   2 marks

Question 12 (3 marks)

A concreter has been asked to concrete a yard based on the plan shown.

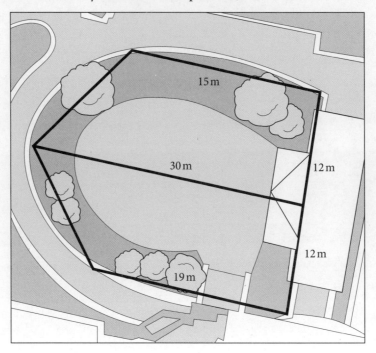

a   Calculate an estimate for the area, using 2 applications of the trapezoidal rule.   1 mark

_____

_____

_____

b   If the concrete is to be 10 cm thick, calculate how many cubic metres of concrete are needed   1 mark
for this area.

_____

_____

_____

c   The cost of the concrete is $180 per cubic metre. The concreter then adds 28% on top of the
cost of the concrete, to cover labour.

How much will the total job cost?   1 mark

_____

_____

_____

_____

Question 13 (2 marks)

An LED TV is rated at 90 W. Electricity is charged at 39 cents/kWh.

If the TV is used 5 hours per day for 6 days a week, how much will it cost to run the TV for a year?   2 marks

Question 14 (4 marks)

The wrist sizes and sleeve lengths of 10 children were measured.

Wrist size (w cm)	16	17	14	16	17	18	15	15	17	19	18
Sleeve length (s cm)	57	59	56	58	58	61	56	57	60	62	60

a   Plot the points from the table of values on the number plane below.   1 mark

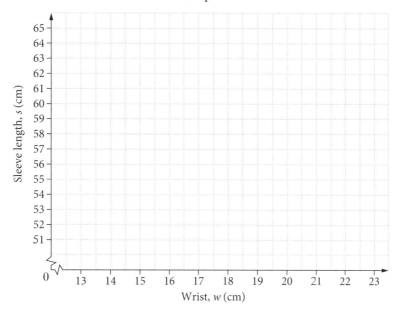

b   Draw a line of best fit by eye for the points plotted.   1 mark

c   Describe the relationship between the two variables and *estimate* Pearson's correlation coefficient.   2 marks

---

Questions 7–14 are worth 18 marks in total (Section II halfway point)

Question 15 (2 marks)

Mina's target heart rate when exercising is calculated using the following formula:

$$\text{THR} = I \times (\text{MHR} - \text{RHR}) + \text{RHR},$$

where I is exercise intensity, MHR is maximum heart rate and RHR is Mina's resting heart rate.

Mina is 42 years old and her maximum heart rate is calculated by using the formula MHR = 220 − her age.

If Mina plans to undertake an exercise of intensity 0.8 and has a resting heart rate of 72 beats per minute, what is her THR, correct to the nearest whole number?   2 marks

Question 16 (4 marks)

The network below shows 11 activities that must be completed to finish a project. The duration of each activity is measured in days.

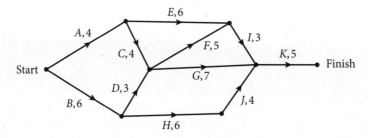

a   How many activities can be delayed without affecting the minimum completion time?   2 marks

b   The project completion time can be lessened by reducing the duration of only 1 activity.

Which activity can be reduced to 1, and what is the new completion time?   2 marks

Question 17 (2 marks)

The number of tilers on a worksite is inversely proportional to the number of days it will take to finish the job. For 2 tilers it would take 17 days to finish the job.

Approximately how many days would 7 tilers take to complete the job?   2 marks

Question 18 (4 marks)

In the following diagram, $\triangle ABC$ has an area of $7\,\text{cm}^2$, $AC = 7\,\text{cm}$, $CD = 3\,\text{cm}$, $\angle BCA = 30°$, $\angle EDB = 55°$, $\angle DBE = 45°11'$ and $\angle BCD = 90°$.

Find the value of x, correct to one decimal place.   4 marks

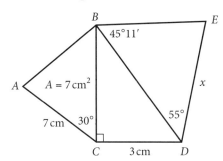

NOT TO SCALE

Question 19 (4 marks)

The table shows monthly repayments of varying interest rates and terms for a loan amount of $100.

Monthly payment amount by loan term (months) and interest rate (%)

	12	24	36	48	60	72	84
2%	$8.42	$4.25	$2.86	$2.17	$1.75	$1.48	$1.28
4%	$8.51	$4.34	$2.95	$2.26	$1.84	$1.56	$1.37
6%	$8.61	$4.43	$3.04	$2.35	$1.93	$1.66	$1.46
8%	$8.70	$4.52	$3.13	$2.44	$2.03	$1.75	$1.56
10%	$8.79	$4.61	$3.23	$2.54	$2.12	$1.85	$1.66
12%	$8.88	$4.71	$3.32	$2.63	$2.22	$1.96	$1.77
14%	$8.98	$4.80	$3.42	$2.73	$2.33	$2.06	$1.87
16%	$9.07	$4.90	$3.52	$2.83	$2.43	$2.17	$1.99

Lana initially thought to borrow $30 000 at 4% p.a. for 6 years. She changes her mind and decides to repay the loan completely in 5 years rather than 6 years.

How much less will she repay in total?   4 marks

END OF PAPER

WORKED SOLUTIONS

Section I (1 mark each)

Question 1

D Vertices B and D have odd degrees (3).

Straightforward question.

Question 2

A −0.8 indicates a strong, negative correlation.

Always look at both direction and strength when estimating the correlation coefficient.

Question 3

B $2 + 4 + 6 = 12$ parts

$$12 \text{ parts} = 180°$$
$$1 \text{ part} = 15°$$
$$2 \text{ parts} = 30°$$

The easiest way to answer this question is to find out what 1 part is equal to. The smallest angle will have the smallest number of parts.

Question 4

A

It is important to know the shapes of non-linear functions. Only showing the graph in the first quadrant is common in HSC questions.

Question 5

B Let x be the required distance.

$$\text{Opposite side} = 3.05 - 1.60$$
$$= 1.45 \text{ m}$$

$$\tan \theta = \frac{\text{opposite}}{\text{adjacent}}$$

$$\tan 21° = \frac{1.45}{x}$$

$$x = \frac{1.45}{\tan 21°}$$

$$= 3.78 \text{ m}$$

Remember that angle of elevation is always above the horizontal line of sight and that it is important to take the height of the line of sight into account.

Question 6

A Future value interest factor for 2.5% for 8 years
$$= 8.7361$$

$$\text{Future value} = 3000 \times 8.7361$$
$$= \$26\,208.30$$

$$\text{Total contributions} = 3000 \times 8$$
$$= \$24\,000$$

$$\text{Interest accrued} = \$26\,208.30 - \$24\,000$$
$$= \$2208.30$$

Finding the right future value interest factor is crucial to answering this type of question. Always read the title of the table because it contains important information. Remember to find the amount of money that was 'contributed' and subtract from the future value to calculate the interest earned.

Section II ($\checkmark$ = 1 mark)

Question 7 (1 mark)

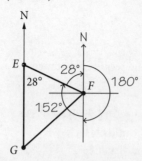

Bearing of E from F = 180° + 152° = 332° $\checkmark$

> Look for alternate angles on parallel lines.
> (Lines pointing north are parallel lines.) This is
> a common question.

Question 8 (2 marks)

a

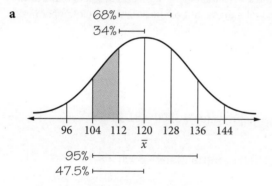

Shaded area: 47.5 – 34 = 13.5% $\checkmark$

b Any score between –1 and –2
(for example, –1.4 or –1.8) $\checkmark$

> Straightforward empirical rule question.
> Understand how to manipulate the percentages to
> isolate certain sections of the normal distribution.

Question 9 (2 marks)

a Line A is $y = -3x - 2$

Line B is $y = 3x - 8$ $\checkmark$

The negative gradient helps to identify which
line represents each equation. Also look at the
y-intercept and that will confirm the correct
equation for each line.

b The solution is the point of intersection of the
two lines. The solution is $x = 1$ and $y = -5$. $\checkmark$

> The coordinates of the point of intersection of a
> pair of lines is always the solution to simultaneous
> equations when graphed. In identifying the correct
> lines, look at the sign of the number in front of
> x. A negative number means the line is sloping
> downwards. A positive number means the line
> is sloping up.

Question 10 (2 marks)

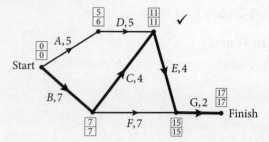

Float time = LST – EST – duration

$= 11 - 5 - 5$

$= 1$ hour $\checkmark$

> Remember that float time is not always
> LST – EST because that vertex may be on the
> critical path.

Question 11 (2 marks)

Maths:

$z = \dfrac{x - \mu}{\sigma}$

$ = \dfrac{71 - 57.8}{6}$

$ = 2.2$

English:

$z = \dfrac{79 - 64.3}{7}$

$ = 2.1$ $\checkmark$

Therefore, Amalia performed better in Maths
as she had a higher z-score in that subject. $\checkmark$

> Common, straightforward question. z-scores are
> the better indicator of test performance and are
> generally used in this context.

Question 12 (3 marks)

a $A = \dfrac{12}{2}(19 + 30) + \dfrac{12}{2}(30 + 15)$

$ = 564 \, \text{m}^2$ $\checkmark$

b $V = Ah$

$ = 564 \times 0.1$

$ = 56.4 \, \text{m}^3$ $\checkmark$

c Cost = $(56.4 \times 180) \times 1.28$

$ = \$12\,994.56$ $\checkmark$

> Volume of a prism ($V = Ah$), as well as associated
> costs, are the logical progression from the
> trapezoidal rule. Make sure that you are familiar
> with this type of question.

9780170459211

Question 13 (2 marks)

$90\,W = 0.09\,kW$

OR

$90\,W$ usage for 1 hour $= 0.09\,kWh$

$\begin{aligned}1 \text{ year's usage} &= 0.09 \times 5 \times 6 \times 52 \\ &= 140.4\,kWh \;\checkmark\end{aligned}$

$\begin{aligned}\text{Electricity cost} &= 140.4 \times 0.39 \\ &= \$54.76 \;\checkmark\end{aligned}$

> Be familiar with the fact that the power usage of an appliance is the number of kilowatts that it will draw with 1 hour of continuous use. Also remember that the appliance's power ratings are measured in watts, meaning you will have to convert to kilowatts before solving.

Question 14 (4 marks)

a

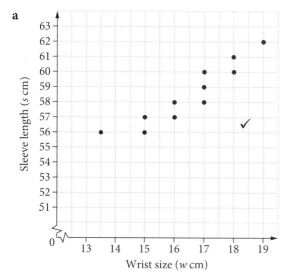

b

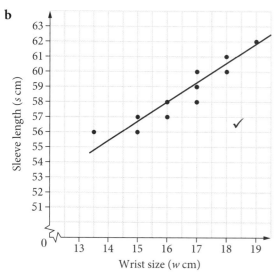

c The relationship is strong, positive and linear. ✓

An estimate of the correlation coefficient could be $r = 0.88$. ✓

Any answer between 0.75 and 0.95 would be accepted.

> Drawing a line of best fit by eye is a necessary skill. Try to have similar numbers of points above and below the line, as well as trying to pass the line through as many points as possible.

Question 15 (2 marks)

$\begin{aligned}\text{MHR} &= 220 - 42 \\ &= 178\,bpm \;\checkmark\end{aligned}$

$\begin{aligned}\text{THR} &= 0.8(178 - 72) + 72 \\ &= 156.8 \\ &\approx 157\,bpm \;\checkmark\end{aligned}$

> These types of questions are straightforward and require you to substitute into formulas and solve correctly.

Question 16 (4 marks)

a To find out how many activities can be delayed, find out which activities are on the critical path.

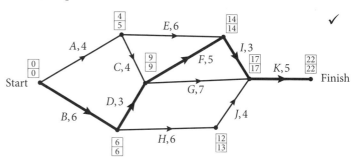

This shows that activities A, C, E, G, H and J are not on the critical path, meaning 6 activities can be delayed without affecting the minimum completion time. ✓

b The first thing to do is write down the length of all paths in this network, from start to finish.

$BDFIK = 22$

$AEIK = 18$

$ACFIK = 21$

$ACGK = 20$

$BDGK = 21$

$BHJK = 21$

When reducing an activity, do not create a new critical path. Reducing activity K will change all paths, meaning $BDFIK$ will still remain the critical path.

Therefore, activity K can be reduced by up to 4 days to leave $K = 1$. ✓

Therefore, the completion time can be reduced to 18 days. ✓

These questions require a good understanding of the importance of the critical path to project completion time. Being able to find the critical path is an important skill to master.

Question 17 (2 marks)

$T = \dfrac{k}{D}$

When $T = 2$, $D = 17$:

$2 = \dfrac{k}{17}$

$k = 34$ ✓

So $T = \dfrac{34}{D}$

$7 = \dfrac{34}{D}$

$D = \dfrac{34}{7}$ days

$= 4.857\ldots$

≈ 5 days ✓

This is a routine question. Always find the value of k in an inverse variation question.

Question 18 (4 marks)

Area of $\triangle ABC = \dfrac{1}{2}ab\sin C$

$7 = \dfrac{1}{2} \times 7 \times BC \times \sin 30°$

$BC = \dfrac{7}{1.75}$

$= 4$ ✓

In $\triangle BCD$:

$c^2 = a^2 + b^2$

$BD^2 = 4^2 + 3^2$

$BD = \sqrt{25}$

$= 5$ ✓

In $\triangle BED$, $180° - 45°11' - 55° = 79°49'$ ✓

So $\dfrac{x}{\sin 45°11'} = \dfrac{5}{\sin 79°49'}$

$x = \dfrac{5}{\sin 79°49'} \times \sin 45°11'$

$= 3.6\,\text{cm}$ ✓

Complex questions worth 4 or 5 marks that are not broken up into parts (with guided clues) are going to become more common. You will need to spend a minute planning your problem-solving steps. You need to develop a logical and systematic approach.

Question 19 (4 marks)

Monthly repayment for 6 years.

Amount at 4% for 72 months is $1.56. ✓

Table is for $100 loans:
$30\,000 \div 100 = \$300$

Monthly repayment for 6 years $= 1.56 \times 300$

$= \$468$ per month

Total paid $= 468 \times 12 \times 6$

$= \$33\,696$ ✓

4% at 60 months = $1.84

Monthly repayment for 5 years

Amount at 4% for 60 months is 1.84.

Monthly repayment $= 1.84 \times 300$

$= \$552$ per month

Total paid $= 552 \times 60$

$= \$33\,120$ ✓

Savings $= \$33\,696 - \$33\,120$

$= \$576$ ✓

Remember that after the monthly repayment is found, it is easy to find out how much in total was paid by multiplying the monthly repayment by the number of months in the term of the loan.

Mathematics Standard 2

PRACTICE MINI-HSC EXAM 2

General instructions	• Reading time: 4 minutes
	• Working time: 1 hour
	• A reference sheet is provided on page 200 at the back of this book
	• For questions in Section II, show relevant mathematical reasoning and/or calculations
Total marks: 40	**Section I – 6 questions, 6 marks**
	• Attempt Questions 1–6
	• Allow about 10 minutes for this section
	Section II – 14 questions, 34 marks
	• Attempt Questions 7–20
	• Allow about 50 minutes for this section

Section I

6 marks
Attempt Questions 1–6
Allow about 10 minutes for this section

Circle the correct answer.

Question 1

What is N 30° W as a true bearing?

A   030°

B   150°

C   300°

D   330°

Question 2

Which graph shows a normal distribution with the highest standard deviation?

A
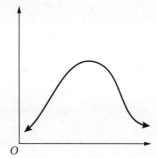

B
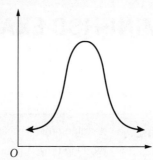

C
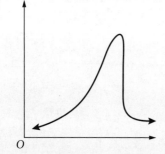

D
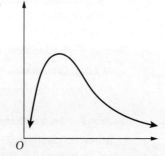

Question 3

A three-course dinner at a restaurant for 2 people cost $95.00 in 2020.

If the average inflation rate is 2.6% p.a., how much would the same meal have cost in 1988?

A $11.80

B $41.78

C $47.80

D $52.40

Question 4

Anh has a large bucket of blue jellybeans. His son adds 72 orange jellybeans and mixes them together with the blue jellybeans. Anh takes out 40 jellybeans and notices that 32 of them are blue.

Approximately how many blue jellybeans are in the bucket?

A   228

B   288

C   320

D   360

Question 5

The graph below represents $y = 1.2^x$. What is the approximate value of $\dfrac{3(1.2)^6}{1.2^{12}}$?

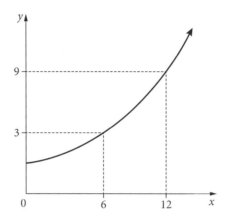

A   1

B   3

C   6

D   9

Question 6

The network below has 6 vertices and 6 edges.

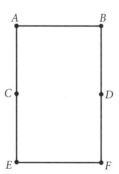

Two new edges are added to the network. Each edge joins 2 of the existing vertices, but not the same 2.

Which of the following CANNOT occur after the 2 edges are added?

A   4 vertices with even degrees

B   1 vertex with a degree of 4

C   All vertices with even degrees

D   4 vertices with odd degrees

Section II

34 marks
Attempt Questions 7–20
Allow about 50 minutes for this section

- Answer the questions in the spaces provided. These spaces provide guidance for the expected length of response.
- Your responses should include relevant mathematical reasoning and/or calculations.

Question 7 (1 mark)

Pina wants to take a long drive from her house in Camden to Grafton, 668 km away. She leaves her house at 12:30 pm and wants to get to Grafton by 8 pm.

At what average speed must she travel to reach Grafton by 8 pm, assuming she does not stop along the way? Give your answer to the nearest whole number.

1 mark

Question 8 (2 marks)

Ebony and Anthony play basketball on the same team. In a particular game, their team scored 84 points. Ebony's and Anthony's points compared to the rest of their team was in the ratio of 6 : 8.

If the ratio of Ebony's points to Anthony's points was 9 : 3, how many points did Anthony score?

2 marks

Question 9 (1 mark)

The graph of the line with equation $y = 2x - 1$ is shown below.

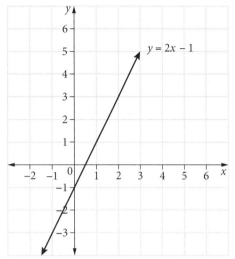

If the graph of $x - y = -1$ is also drawn on this number plane, what is the point of intersection of the two lines?   1 mark

Question 10 (2 marks)

A machine packs chips into bags. Each bag is weighed and the results are normally distributed. When a total of 5 bags are 3 or more standard deviations from the mean, the machine is shut down and recalibrated.

If the machine produces 5000 bags a day, how many times would you expect it to be recalibrated in a day?   2 marks

Question 11 (1 mark)

Draw a scatterplot of 5 points that demonstrates a correlation coefficient value of −1.   1 mark

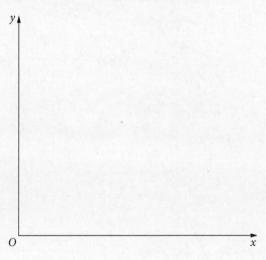

Question 12 (2 marks)

Con has a car that can travel 480 km on a 70-litre tank of petrol. He decides to buy a more fuel-efficient car. The salesperson tells him the new car's fuel consumption, in litres per kilometre, is 35% less than that of his current car.

What is the fuel consumption rate of the new car? Give your answer correct to one decimal place.   2 marks

Question 13 (1 mark)

A certain project has 4 activities that must be completed. The table below shows the duration of each activity in days, and the immediate predecessor for each activity.

Activity	Duration (days)	Immediate predecessor
A	4	–
B	10	A
C	3	B
D	14	C

What is the latest starting time for activity *D* if the whole project needs to be completed in 34 days?   1 mark

Question 14 (4 marks)

A researcher is investigating how radioactive decay affects a block of radioactive material. The decay of the material can be modelled by the formula $M = 1000(0.7)^n$, where M is mass of the block in grams and n is number of years.

a What was the initial mass of the block? 1 mark

b By what percentage is the block decaying each year? 1 mark

c Estimate the mass of the block after 7 years, correct to one decimal place. 1 mark

d How long (to the nearest year) will the block take to have a mass less than 20 g? 1 mark

Question 15 (3 marks)

A triangle has sides of lengths 28 m, 31 m, x m and an angle of 73°.

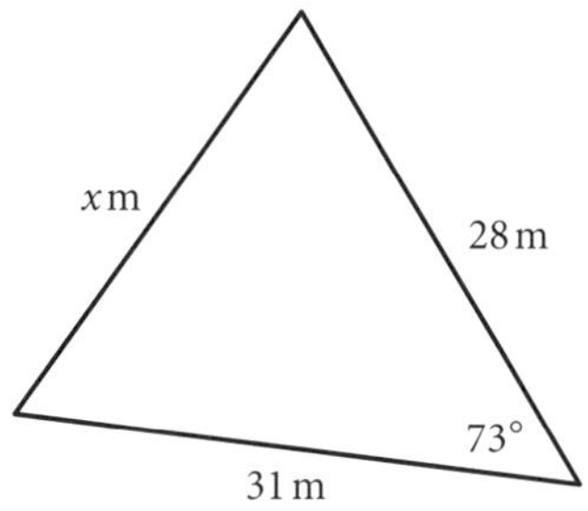

NOT TO SCALE

x m

28 m

73°

31 m

What is the value of x, correct to three significant figures? 3 marks

9780170459211

Question 16 (3 marks)

Manny has a credit card with the following features:

- No interest-free period

- Compound interest charged at 0.025% per day

- Minimum payment of $50 or 5% of the balance outstanding (whichever is larger) at each statement date

If Manny purchases 5 pairs of boxing gloves for $1040, what is the minimum payment he must make on the   3 marks
statement date if the balance is outstanding for 42 days? Show your answer with appropriate calculations.

Question 17 (2 marks)

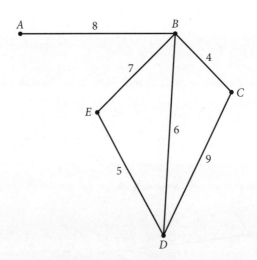

Complete the table below to represent the weighted network shown.   2 marks

	A	B	C	D	E
A	–	8	0		0
B		–			
C			–		
D				–	5
E				5	–

Question 18 (5 marks)

Sue travels by air for 870 km from B to C on a bearing of 040°. She travels south-east from C to D for 942 km. The bearing of B from D is 272°.

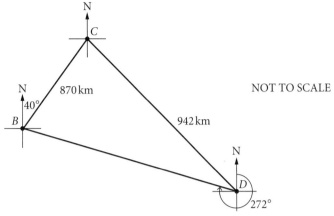

If Sue travels back to B at a speed of 450 km/h, how long will her flight take, in hours and minutes, and what is the three-figure bearing of D from B?   5 marks

Question 19 (3 marks)

The network below shows water pipes connecting various locations around a city. The water flows from the source to the sink and the numbers represent the capacities of the pipes, in kilolitres per hour.

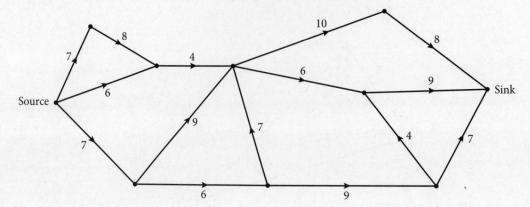

a What is the maximum capacity of this network?   1 mark

_____

_____

_____

b If all of the pipes with a capacity of 7 were upgraded to a capacity of 10, by how much would   2 marks
the new maximum flow increase?

_____

_____

_____

_____

_____

_____

_____

_____

Question 20 (4 marks)

Adrian thinks he can predict the height of a person by the length of their stride (pace) when they walk.
He measured the heights of 10 students and the lengths of their strides.

Stride (s cm)	Height (h cm)
103	162
82	138
107	181
108	177
97	160
96	165
104	177
100	176
90	147
94	155

a   For this data, find the equation of the least-squares regression line for h. Give your answers correct   2 marks
to four decimal places.

b   What is a limitation of the model in predicting height?   1 mark

c   Predict the stride length of a person with a height of 169 cm, correct to the nearest centimetre.   1 mark

END OF PAPER

9780170459211

SECTION II EXTRA WRITING SPACE

WORKED SOLUTIONS

Section I (1 mark each)

Question 1

D

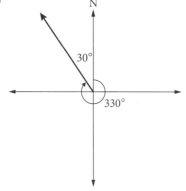

Straightforward question. Although three-figure bearings are more common in questions, it is important to know how compass bearings work.

Question 2

A

Normal distribution means the graph is symmetrical. A high standard deviation means the scores are further spread apart.

Question 3

B $FV = PV(1 + r)^n$

$95 = PV(1 + 0.026)^{32}$

$PV = \dfrac{95}{(1.026)^{32}}$

$= \$41.78$

Remember that inflation and appreciation work in much the same way as compound interest.

Question 4

B Let x be the total number of jellybeans in the bucket.

$\dfrac{x}{72} = \dfrac{40}{8}$

$x = \dfrac{40}{8} \times 72$

$= 360$ total jellybeans

Blue jellybeans $= 360 - 72$

$= 288$

Capture–recapture is a straightforward concept. Check to make sure you answer the exact question being asked.

Question 5

A $1.2^6 = 3$ and $1.2^{12} = 9$

So $\dfrac{3(3)}{9} = 1$

This was a different substitution question. Being able to interpret non-linear graphs is an important skill.

Question 6

C Before the 2 edges are added, all vertices have a degree of 2. If the 2 new edges join 2 vertices each, then 3 or 4 vertices will have degree increases. If 1 vertex is shared, then its degree is 4, so 2 odd, 4 even. Otherwise, 4 vertices have degree 3, so 4 odd, 2 even.

Questions such as this require experimentation. This may take some time; however, the solutions are usually limited to a few combinations.

Section II (✓ = 1 mark)

Question 7 (1 mark)

$$S = \frac{D}{T}$$

$$= \frac{668}{7.5}$$

$$= 89\,\text{km/h} \checkmark$$

> Straightforward question. If needed, refer to the *DST* triangle to help you find any of the three variables.

Question 8 (2 marks)

Total parts = 6 + 8 = 14

14 parts = 84 points

1 part = 6 points

So Ebony and Anthony = 36 points ✓

Ebony and Anthony total parts = 12

12 parts = 36 points

1 part = 3 points

So Anthony = 9 points ✓

> In ratios, always find out what 1 part is equal to. These questions should be achievable with practice.

Question 9 (1 mark)

Point of intersection will be $x = 2$, $y = 3$ ✓

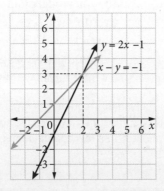

> Make sure that you can sketch linear graphs quickly using the intercept method. The point of intersection of the two lines is also the solution of the 2 simultaneous equations.

Question 10 (2 marks)

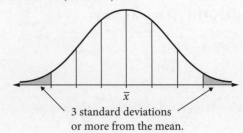

3 standard deviations or more from the mean.

Shaded area = 100 − 99.7 = 0.3%

So 5000 × 0.003 = 15 bags ✓

So number of times to recalibrate

$$= \frac{15}{5}$$

$$\approx 3 \text{ times per day} \checkmark$$

> The percentages for a normal distribution can also be applied to probability and expectation questions.

Question 11 (1 mark)

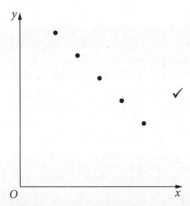

> Points with a correlation coefficient of −1 are in a perfectly straight line that slopes downwards.

Question 12 (2 marks)

Old car fuel consumption $= \dfrac{70}{4.8}$

$$= 14.6\,\text{L/100\,km} \checkmark$$

New car is 35% more efficient, so fuel consumption is 65% of the old car.

New car fuel consumption = 14.6 × 0.65

$$= 9.5\,\text{L/100\,km} \checkmark$$

> Remember that fuel consumption is measured in L/100 km. Always try to get the denominator to 100.

9780170459211

Question 13 (1 mark)

Each activity follows on from another. Activity D has a duration of 14 days. The entire project is to be completed in 34 days.

Therefore, latest starting time for activity D

$= 34 - 14$

$= 20$ days ✓

> When looking at float times, always look at the duration of activities.

Question 14 (4 marks)

a $1000\,g = 1\,kg$ (value of M when $n = 0$) ✓

b $100 - 70 = 30\%$ per year ✓

(0.7 refers to 70% of the original value, thereby losing 30% each time period.)

c $M = 1000(0.7)^7 = 82.4\,g$ ✓

d $n = 10$ years, $M = 28\,g$
$n = 11$ years, $M = 19.77\,g$

So it takes 11 years for the block's mass to be less than 20 g. ✓

> Trial and error is the usual way to find values of n; however, if you have experience using logarithms, you can find that method helpful as well.

Question 15 (3 marks)

$c^2 = a^2 + b^2 - 2ab\cos C$
$x^2 = 28^2 + 31^2 - 2 \times 28 \times 31 \times \cos 73°$ ✓
$x = \sqrt{1237.44\ldots}$
$\quad = 35.1773$ ✓
$\quad \approx 35.2\,m$ ✓

> Straightforward question. Remember that the sine rule is 2 sides, 2 angles and the cosine rule is 3 sides, 1 angle. In this question, 1 mark was allocated for correct rounding. Always pay attention to rounding in all of the questions that you answer.

Question 16 (3 marks)

$A = 1040(1 + 0.000\,25)^{42}$ ✓
$\quad = \$1050.98$

5% of $1050.98 = \$52.55$ ✓

This is greater than $50 so minimum payment
$= \$52.55$ ✓

> Credit card questions present a lot of opportunities to make mistakes. Be careful with the daily interest rate and make sure that you divide that rate by 100 to make it a decimal. Also, make sure that you write a concluding sentence that backs up your calculations and answers the question clearly.

Question 17 (2 marks)

	A	B	C	D	E	
A	–	8	0	0	0	
B	8	–	4	6	7	✓
C	0	4	–	9	0	
D	0	6	9	–	5	✓
E	0	7	0	5	–	

> Reading a network from a table and vice-versa is an important skill in the topic of networks.

Question 18 (5 marks)

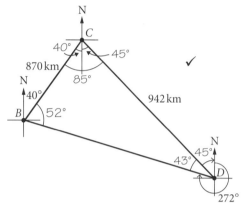

$c^2 = a^2 + b^2 - 2ab\cos C$
$BD^2 = 870^2 + 942^2 - 2 \times 870 \times 942 \times \cos 85°$ ✓
$BD = \sqrt{1\,501\,408.765}$
$\quad = 1225.32\ldots$ ✓

$\text{Time} = \dfrac{1225.32\ldots}{450 \text{ km/h}}$
$\quad = 2.7229\ldots$
$\quad = 2$ hours and 43 minutes ✓

$\angle CDB = 360° - 272° - 45° = 43°$
$\angle CBD = 180° - 85° - 43° = 52°$

Bearing of D from $B = 40° + 52° = 092°$ ✓

> Questions worth 4 or 5 marks that are not broken up into parts are going to become more common. With trigonometry questions, always work through all of the angles in the diagram. This will help you to piece together the answer required and check that all parts of the question are answered.

Question 19 (3 marks)

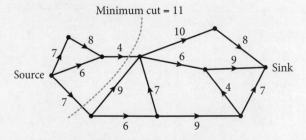

Minimum cut = 11

a The minimum cut is also the maximum capacity of the network. Therefore, the maximum capacity is 11 kL/h. ✓

b

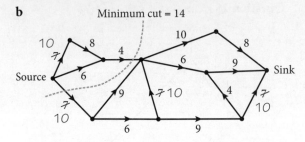

Minimum cut = 14

An upgrade of 7 to 10 will mean the minimum cut will now be 14. ✓

This means there will be an increase of 3 kL/h in the maximum capacity of the network. ✓

> Remember that the minimum cut will be equal to the maximum flow of the network. The maximum flow will only increase if the edges on the minimum cut are increased.

Question 20 (4 marks)

a From calculator:

$c = 2.4674\ldots$ and $m = 1.6446\ldots$ ✓

So $h = 1.6446s + 2.4674$ ✓

b Human strides and heights are limited by a physical range. There is a limit to height because people eventually stop growing. ✓

c
$$h = 1.6446s + 2.4674$$
$$169 = 1.6446s + 2.4674$$
$$s = \frac{166.533}{1.6446}$$
$$= 101.26 \text{ cm}$$
$$\approx 101 \text{ cm} ✓$$

> It is important to know how to find the equation of the least-squares regression line from your calculator. Knowing '$y = bx + a$' on your calculator and what that means will make questions like this much easier.

Mathematics Standard 2

PRACTICE HSC EXAM 1

General instructions	• Reading time: 10 minutes
	• Working time: 2 hours and 30 minutes
	• A reference sheet is provided on page 200 at the back of this book
	• For questions in Section II, show relevant mathematical reasoning and/or calculations

Total marks: 100

Section I – 15 questions, 15 marks

• Attempt Questions 1–15

• Allow about 25 minutes for this section

Section II – 30 questions, 85 marks

• Attempt Questions 16–45

• Allow about 2 hours and 5 minutes for this section

Section I

15 marks
Attempt Questions 1–15
Allow about 25 minutes for this section

Circle the correct answer.

Question 1

Which of the following is NOT a path for the network below?

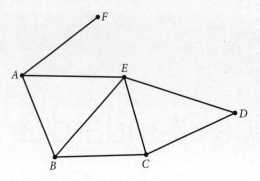

A   FAEBC

B   BCDEAF

C   DCBAEF

D   CDEBAF

Question 2

What is 34 478 900 written in scientific notation, correct to three significant figures?

A 3.4×10^{7}

B 3.44×10^{7}

C 3.45×10^{7}

D 34.5×10^{6}

Question 3

Po has a $\dfrac{7}{13}$ chance of beating Richard in a game of cards. Po and Richard play 2 games.

What is the probability that Po will win at least 1 game against Richard?

A $\dfrac{36}{169}$

B $\dfrac{42}{169}$

C $\dfrac{49}{169}$

D $\dfrac{133}{169}$

Question 4

Young's formula is used to calculate the medicine dosage for children aged 1–12 years based on the adult dosage. The formula is:

$$D = \frac{yA}{y + 12},$$

where D = dosage amount
y = age of child, in years
A = adult dosage.

If the adult dosage for a particular medicine is 60 mg and the child's dosage is 20 mg, how old is the child in years?

A   6

B   7

C   8

D   9

Question 5

Which description best describes the relationship shown in the scatterplot below?

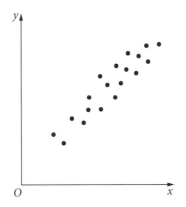

A   Linear, negative, weak

B   Non-linear, positive, strong

C   Non-linear, negative, weak

D   Linear, positive, strong

Question 6

Jay visits his local garden centre to buy cement blocks for his garden. The charge for each block is shown in the table below.

Number of blocks (B)	1	2	3	4
Cost (C)	15.71	30.01	44.31	58.61

Which linear equation shows the relationship between blocks (B) and cost (C)?

A $C = 14.3B + 1.41$

B $C = 14.3B + 2.41$

C $C = 15.71B + 1.41$

D $C = 15.71B - 1.41$

Question 7

The time taken to drive to a destination varies inversely with the speed of a car.

Which graph best represents this situation?

A
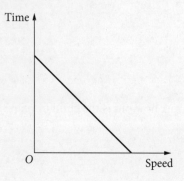

B
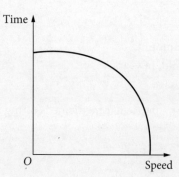

C
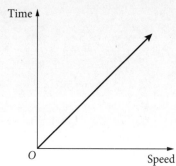

D
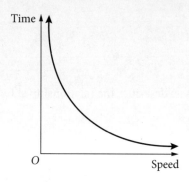

Question 8

Which equation shows f as the subject of $a = cd + \dfrac{f}{k}$?

A $f = \dfrac{a + cd}{k}$

B $f = ak - kcd$

C $f = \dfrac{a}{cd} - k$

D $f = \dfrac{k - a}{cd}$

Question 9

What is the distance from A to B in the diagram below?

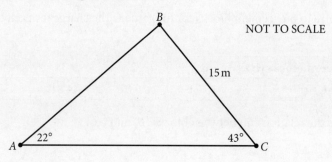

NOT TO SCALE

A   17 m

B   20 m

C   24 m

D   27 m

Question 10

Electricity prices have risen by 3.4% p.a. over the past few years.

If the average electricity bill in 2012 was $425, what was the average electricity bill in 2020?

A $437.75

B $489.22

C $523.67

D $555.33

Question 11

Ravi achieved a z-score of 2.4 on his latest Biology test.

If the test scores were normally distributed with a mean of 56 and a standard deviation of 5, what was Ravi's actual test mark?

A   58

B   65

C   67

D   68

Question 12

Sydney City Council has announced a plan to renovate and redesign the area around the Hyde Park fountain. The council has asked you to estimate the size of the area outlined below.

Using 2 applications of the trapezoidal rule, what is the estimated area of redevelopment?

A $28\,680\,\text{m}^2$

B $32\,460\,\text{m}^2$

C $32\,750\,\text{m}^2$

D $34\,750\,\text{m}^2$

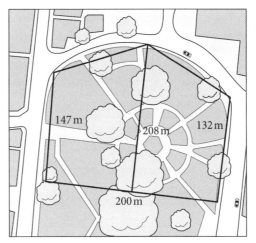

NOT TO SCALE

Question 13

Scientists were observing a group of 830 penguins in Antarctica. The group consisted of 372 emperor penguins, and the rest were rockhopper penguins.

What is the ratio of emperor penguins to rockhopper penguins, in fully simplified form?

A   372:258

B   372:830

C   186:229

D   186:415

Question 14

If the area of the triangle below is 23.09 cm², what is the size of $\angle ACB$, rounded to the nearest degree?

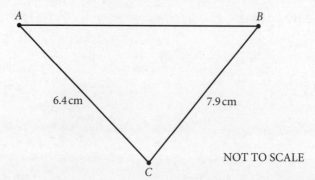

NOT TO SCALE

A   24°

B   54°

C   65°

D   66°

Question 15

Ellie needs a $60 000 deposit in order to buy her first home. She would like to have her deposit ready in 5 years. She finds an annuity that will pay 4% p.a. and she plans to contribute money every 6 months.

With reference to the table below, how much will Ellie need to contribute each time period in order to save her deposit of $60 000 in 5 years?

Future value interest factors
(Future value of an annuity with a contribution of $1 at the end of each period)

Period	Interest rate per period									
	0.25%	0.50%	0.75%	1.00%	1.50%	2.00%	2.50%	3.00%	4.00%	5.00%
1	1.0000	1.0000	1.0000	1.0000	1.0000	1.0000	1.0000	1.0000	1.0000	1.0000
2	2.0025	2.0050	2.0075	2.0100	2.0150	2.0200	2.0250	2.0300	2.0400	2.0500
3	3.0075	3.0150	3.0226	3.0301	3.0452	3.0604	3.0756	3.0909	3.1216	3.1525
4	4.0150	4.0301	4.0452	4.0604	4.0909	4.1216	4.1525	4.1836	4.2465	4.3101
5	5.0251	5.0503	5.0756	5.1010	5.1523	5.2040	5.2563	5.3091	5.4163	5.5256
6	6.0376	6.0755	6.1136	6.1520	6.2296	6.3081	6.3877	6.4684	6.6330	6.8019
7	7.0527	7.1059	7.1595	7.2135	7.3230	7.4343	7.5474	7.6625	7.8983	8.1420
8	8.0704	8.1414	8.2132	8.2857	8.4328	8.5830	8.7361	8.8923	9.2142	9.5491
9	9.0905	9.1821	9.2748	9.3685	9.5593	9.7546	9.9545	10.1591	10.5828	11.0266
10	10.1133	10.2280	10.3443	10.4622	10.7027	10.9497	11.2034	11.4639	12.0061	12.5779
11	11.1385	11.2792	11.4219	11.5668	11.8633	12.1687	12.4835	12.8078	13.4864	14.2068
12	12.1664	12.3356	12.5076	12.6825	13.0412	13.4121	13.7956	14.1920	15.0258	15.9171

A $4997.46

B $5479.60

C   11 077.67

D $11 529

Section II

85 marks
Attempt Questions 16–45
Allow about 2 hours and 5 minutes for this section

- Answer the questions in the spaces provided. These spaces provide guidance for the expected length of response.
- Your responses should include relevant mathematical reasoning and/or calculations.

Question 16 (1 mark)

The following symbol is found on a house plan.

What does the symbol represent?   1 mark

Question 17 (2 marks)

A bread recipe uses 42 g of sugar, a small amount of salt and a large amount of flour.

If the ratio of sugar to salt to flour is $3:1:55$, how much flour needs to be added to make a loaf of bread?   2 marks

Question 18 (1 mark)

A vacuum cleaner has the following energy label on it. The vacuum cleaner is rated at 2000 W of power.

How many hours and minutes of usage per week is assumed, given the energy label? Give your answer   1 mark
to the nearest minute.

Question 19 (2 marks)

Lourdes needs $14 000 for her Europe trip of a lifetime. She is planning to go in 5 years' time. She decides to invest an amount of money into an account that pays 3.6% p.a. interest compounded monthly.

How much should Lourdes invest now to ensure she has the funds for her Europe trip?   2 marks

Question 20 (2 marks)

A 2 kg packet of sugar costs $2.79. A 600 g packet of sugar costs $0.90.

What is the difference in price between the packets, per 500 g?   2 marks

Question 21 (3 marks)

An outdoor lounge set was purchased for $3800 using a credit card. The credit card has interest compounded daily at the rate of 22.1% p.a. There is no interest-free period.

If the outstanding amount is paid in full after 26 days, how much interest is paid?   3 marks

Question 22 (4 marks)

Jennifer has a net income of $920 per week and prepares the following weekly budget.

Rent	$280
Petrol	$50
Other car costs	$25
Food	$180
Utilities	$25
Mobile phone	$20
Insurance	$120
Entertainment	$100
Savings	?
	$920

a What percentage of her income does she plan to save? Round your answer to the nearest whole number. 2 marks

b If the cost of all items apart from Savings increases by 5%, how much will she be able to put aside for Savings? 2 marks

Question 23 (2 marks)

The blood alcohol content (BAC) for males is calculated using the following formula:

$$BAC_{male} = \frac{10N - 7.5H}{6.8M},$$

where N = number of standard drinks

H = number of hours spent drinking

M = person's mass in kilograms.

Hunter weighs 87 kg and wants to stay under the legal limit of 0.05.

How many standard drinks can he consume over 3 hours without going over the limit? 2 marks

Question 24 (2 marks)

Lani takes a 14-hour flight from Edmonton, Canada (UTC –7) to Paris, France (UTC +5). She leaves at 9 am on a Monday morning.

What is the time in Paris when she arrives? 2 marks

Question 25 (4 marks)

A project has 12 specific activities that need to take place for it to be completed. The table below shows activities, duration (days) and immediate predecessors.

Activity	Duration (days)	Immediate predecessor(s)
A	3	–
B	4	–
C	3	–
D	2	C
E	3	C
F	4	E
G	1	B, D, F
H	3	G
I	4	A
J	7	H, I
K	8	G
L	9	J, K

The network depicting the table is shown below, with 3 activities missing.

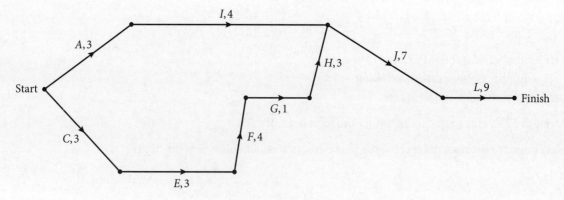

a Draw in the 3 activities that are missing in the network above and include labels and durations. 2 marks

b What is the minimum completion time of the project? 2 marks

Question 26 (3 marks)

Jessica has 1500 shares. The market price is $4.30 per share. Jessica is paid a dividend of $0.40 per share.

a   What is the dividend yield of Jessica's shares, correct to one decimal place?   1 mark

b   Assuming the market price stays the same, by how much would the dividend increase if the dividend   2 marks
yield was 3.4% higher the following year?

Question 27 (2 marks)

Dale works out every day and has a resting heart rate of 70 beats per minute. During a 4-minute, high-intensity workout, he gets his heart rate to beat at 205% of his resting heart rate for the full 4 minutes.

How many times does his heart beat during the 4-minute workout?   2 marks

Question 28 (3 marks)

On a map, the scaled distance from Camp A to Camp B is 3.46 cm but in reality they are 17.3 km apart.
The distance from Camp B to Camp C is 24.3 km.

What is the scaled distance between Camp B and Camp C on the map?   3 marks

Question 29 (2 marks)

Alison is car shopping and pays close attention to fuel consumption rates. She notices that one car has a rate of 10.2 L/100 km for city driving and 8.1 L/100 km for highway driving.

She buys the car and takes it on a 680 km trip, of which about 430 km was highway driving and the rest was city driving.

Approximately how much petrol did Alison's car consume for the journey?   2 marks

Question 30 (4 marks)

The network below shows the activities needed to finish a particular project as well as their durations, measured in days.

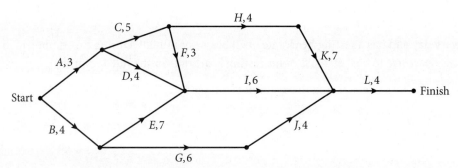

a   What is the minimum completion time for the project?   1 mark

b   How many activities have a float time of 0?   2 marks

c   If the duration of activity *I* was changed from 6 to 8 days, how many more activities would have   1 mark
a float time of 0?

Question 31 (4 marks)

For a set of rectangles, the relationship between area, A, and width, w, is given by $A = w^2 + 4w$.
The graph of $A = w^2 + 4w$ is shown below.

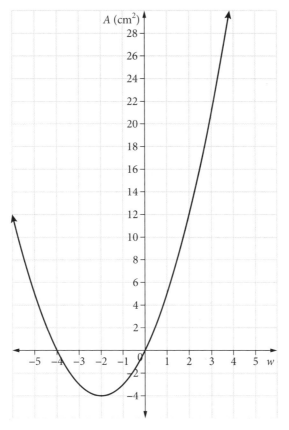

a   If a rectangle's width is 1 cm, what is its area?   1 mark

_____

b   If a rectangle's width is 2 cm, what is its length?   1 mark

_____

_____

_____

c   What is the difference in perimeter between a rectangle with a width of 2 cm and a rectangle with   2 marks
a width of 3 cm?

_____

_____

_____

_____

_____

Question 32 (3 marks)

The time taken, T (hours), for a car to travel between Town A and Town B varies inversely with the average speed, S (km/h).

A car that leaves Town A at 8 am at an average speed of 80 km/h will arrive in Town B at 5 pm.

At what time will the car arrive in Town B if it travels at an average speed of 90 km/h?   3 marks

_____

_____

_____

_____

_____

_____

Questions 16–32 are worth 44 marks in total (Section II halfway point)

Question 33 (3 marks)

The weighted network below shows the connections between 8 areas.

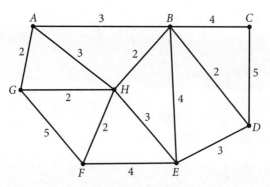

a   Draw the 2 minimum spanning trees from the network.   2 marks

b   What is the length of the minimum spanning tree?   1 mark

_____

_____

Question 34 (3 marks)

The following data set shows the marks of 10 students on a quiz.

2   2   3   4   5   5   5   6   8   9

Draw a box plot for the data below and find the interquartile range.   3 marks

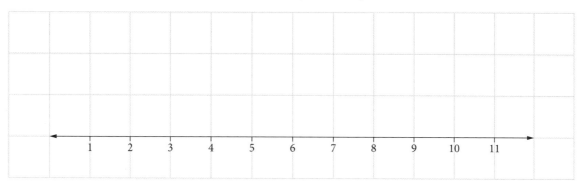

Question 35 (2 marks)

A diagram of a tunnel in the shape of half an annulus is shown below. The tunnel is to be made of concrete.

What is the volume of concrete needed to make the tunnel, to the nearest cubic metre?   2 marks

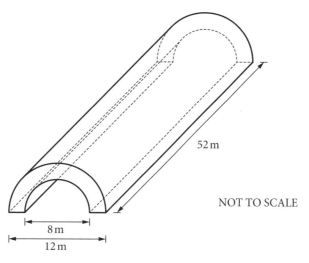

52 m

NOT TO SCALE

8 m

12 m

Question 36 (3 marks)

Katelyn plans to invest $300 per month for 3 years. The account pays 3% per month.

Using the present value interest factor table, calculate the future value of the annuity Katelyn wants   3 marks
to invest in.

Present value of an annuity of $1 per period for n periods

	Interest rate per period									
Period	**1%**	**2%**	**3%**	**4%**	**5%**	**6%**	**7%**	**8%**	**9%**	**10%**
30	25.8077	22.3965	19.6004	17.2920	15.3725	13.7648	12.4090	11.2578	10.2737	9.4269
31	26.5423	22.9377	20.0004	17.5885	15.5928	13.9291	12.5318	11.3498	10.3428	9.4790
32	27.2696	23.4683	20.3888	17.8736	15.8027	14.0840	12.6466	11.4350	10.4062	9.5264
33	27.9897	23.9886	20.7658	18.1476	16.0025	14.2302	12.7538	11.5139	10.4644	9.5694
34	28.7027	24.4986	21.1318	18.4112	16.1929	14.3681	12.8540	11.5869	10.5178	9.6086
35	29.4086	24.9986	21.4872	18.6646	16.3742	14.4982	12.9477	11.6546	10.5668	9.6442
36	30.1075	25.4888	21.8323	18.9083	16.5469	14.6210	13.0352	11.7172	10.6118	9.6765
37	30.7995	25.9695	22.1672	19.1426	16.7113	14.7368	13.1170	11.7752	10.6530	9.7059
38	31.4847	26.4406	22.4925	19.3679	16.8679	14.8460	13.1935	11.8289	10.6908	9.7327
39	32.1630	26.9026	22.8082	19.5845	17.0170	14.9491	13.2649	11.8786	10.7255	9.7570
40	32.8347	27.3555	23.1148	19.7928	17.1591	15.0463	13.3317	11.9246	10.7574	9.7791
41	33.4997	27.7995	23.4124	19.9931	17.2944	15.1380	13.3941	11.9672	10.7866	9.7991

Question 37 (3 marks)

On the number plane below, graph the simultaneous equations $y = -2x + 4$ and $y = \dfrac{1}{4}x - 5$ and hence, solve the equations.

3 marks

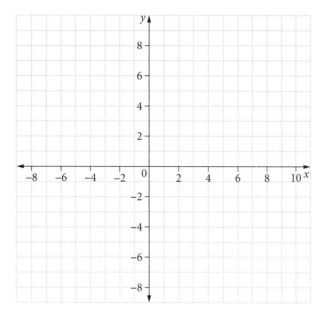

Question 38 (2 marks)

The arm lengths of a group of 20 000 people were normally distributed with a mean of 70 cm and a standard deviation of 3 cm.

How many people had arm lengths between 73 cm and 79 cm?

2 marks

Question 39 (3 marks)

A group of 8 males were studied to see if there was a relationship between their shoe sizes and the lengths of their femurs (thigh bones). The scatterplot below shows the results of the study.

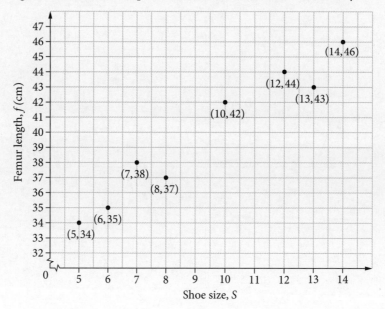

By using the plotted points, find the least-squares regression line (answer correct to three decimal places) and use it to predict the shoe size of a male with a femur length of 40 cm.   **3 marks**

Question 40 (4 marks)

Krystal has a box containing 5 green lollies and 8 orange lollies. She picks 1 lolly at random out of the box and puts it on a table. She then picks a second lolly and puts it next to the first lolly.

a   Complete the tree diagram below and write the correct probabilities on the branches.   3 marks

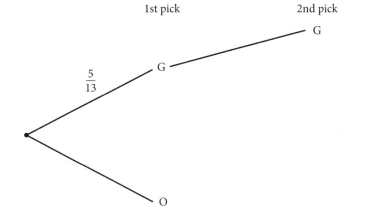

1st pick   2nd pick   Outcomes

G   G G

$\frac{5}{13}$ G

O

b   What is the probability of choosing 1 lolly of each colour?   1 mark

Question 41 (3 marks)

Brianna buys her dream car for $28 950. The car depreciates at a rate of 18% p.a. each year.

a   What is the value of the car in 3 years?   1 mark

b   Approximately how many years (to one decimal place) will the car's value take to decrease to less than $10 000?   2 marks

Question 42 (4 marks)

Harry wants to study the growth in a wallaby population in a certain bushland area. On his first visit to the area, he counts 100 wallabies. Over the next few years, he finds that the wallaby population grows by 5% each year from his first visit. Harry models the growth in the wallaby population with the formula $P = 100(1.05)^n$, where P = population and n = number of years.

a   What was the difference in the wallaby population between the 4th and 8th years?   1 mark

_____

_____

_____

b   Complete the following table and use the points to draw the graph of $P = 100(1.05)^n$. Round values   2 marks
of P to the nearest whole number.

n	2	7	12	17
P	110	141	____	____

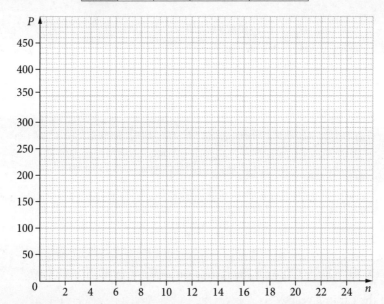

c   Using your graph or otherwise, find (to the nearest year) when the wallaby population will double.   1 mark

_____

_____

Question 43 (3 marks)

An art gallery has different exhibits showing at the same time. The owners want to analyse how many people can move through the gallery and between the exhibits. The edges show the maximum number of people that can walk from one exhibit to another.

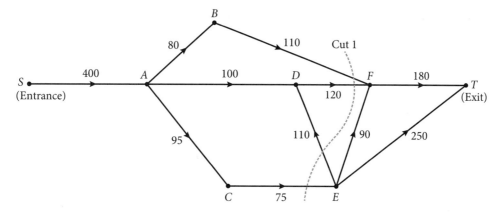

a What is the capacity of cut 1?   1 mark

_____

_____

b Calculate the maximum flow and draw the minimum cut.   2 mark

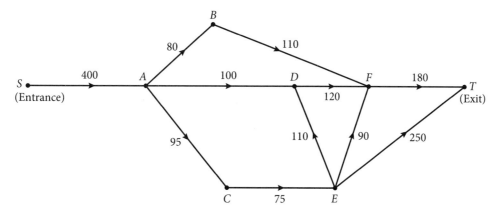

_____

_____

Question 44 (3 marks)

Xan takes out a loan for $300 000 and agrees to pay $1422.63 per month for 25 years. After 3 years, he can afford to pay an extra $160 per month as well as his normal monthly repayments. As a result, he fully repays his loan 3 years early.

How much interest did he save by paying an extra $160 per month after the 3rd year?   3 marks

Question 45 (5 marks)

Tyrone is taking part in a cycling race. He cycles from Q to R on a bearing of 145°. From R, he cycles 8.2 km on a bearing of 230° from S. He then cycles 9.7 km back to Q.

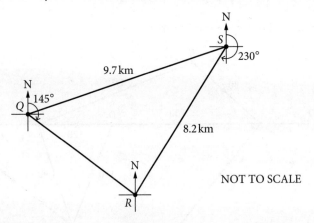

NOT TO SCALE

a   What is the bearing of S from Q, to the nearest degree?   2 marks

b   What is the total distance of the cycling course, correct to one decimal place?   3 marks

END OF PAPER

SECTION II EXTRA WRITING SPACE

WORKED SOLUTIONS

Section I (1 mark each)

Question 1

C

> Straightforward question. The only path that has a disconnect is C, where F and E are not joined directly.

Question 2

C

> Common question to combine scientific (or standard) notation with significant figures. Remember that the decimal point is always moved after the first non-zero number.

Question 3

D $P(\text{at least 1 win}) = 1 - P(\text{no wins})$

$$= 1 - \left(\frac{6}{13} \times \frac{6}{13}\right)$$

$$= \frac{133}{169}$$

> The common phrase 'at least' is easily handled by calculating 1 − probability of none.

Question 4

A $D = \dfrac{yA}{y + 12}$

$$20 = \frac{60y}{y + 12}$$

$$20(y + 12) = 60y$$

$$20y + 240 = 60y$$

$$240 = 40y$$

$$y = 6 \text{ years}$$

> In any algebraic formula, you can be asked to find any variable in that formula. Sharpening your algebraic skills is a good use of study time.

Question 5

D   The points are trending upwards; therefore, it is positive. The points are packed closely together, so it is a strong relationship.

> Correlation looks at direction and strength.

Question 6

A $C = mB + c$

$$m = \frac{y_2 - y_1}{x_2 - x_1}$$

$$= \frac{30.01 - 15.71}{2 - 1}$$

$$= 14.3$$

$C = 14.3B + c$

Substitute any point into the equation, such as $(1, 15.71)$:

$$15.71 = 14.3(1) + c$$

$$c = 15.71 - 14.3$$

$$= 1.41$$

So $C = 14.3B + 1.41$

> Common question. Remember to find both m and c to obtain the linear equation from the table.

Question 7

D

> Inverse variation graphs are hyperbolas.

Question 8

B $a = cd + \dfrac{f}{k}$

$$a - cd = \frac{f}{k}$$

$$k(a - cd) = f$$

$$f = ka - kcd$$

> Changing the subject requires a lot of practice at manipulating algebraic equations. Remember to perform the opposite operation to move a term from one side of the equation to the other.

Question 9

D $\dfrac{a}{\sin A} = \dfrac{b}{\sin B}$

$$\frac{AB}{\sin 43°} = \frac{15}{\sin 22°}$$

$$AB = \frac{15 \sin 43°}{\sin 22°}$$

$$\approx 27 \, \text{m}$$

> Routine sine rule question. Make sure to pair up each side with the angle opposite it.

Question 10

D $FV = PV(1 + r)^n$

$\quad\quad = 425(1 + 0.034)^8$

$\quad\quad = \$555.33$

Inflation questions work exactly the same way as compound interest questions. Make sure that the interest rates and time periods match.

Question 11

D $\quad z = \dfrac{x - \mu}{\sigma}$

$\quad 2.4 = \dfrac{x - 56}{5}$

$\quad 12 = x - 56$

$\quad\quad x = 68$

Straightforward question. Be careful substituting the numbers and take care when rearranging the equation to find x.

Question 12

B $A \approx \dfrac{100}{2}(147 + 208) + \dfrac{100}{2} + (208 + 132)$

$\quad\quad \approx 34\,750\,\text{m}^2$

Routine trapezoidal rule question. Remember that the perpendicular width of each strip should be exactly the same.

Question 13

C Rockhopper = 830 − 372

$\quad\quad\quad\quad\quad = 458$

So emperor : rockhopper

$\quad\quad\quad 372 : 458$

$\quad\quad\quad 186 : 229$

Straightforward ratio question. Remember that the order of the groups in the question is the order of the ratio.

Question 14

D $\quad A = \dfrac{1}{2}ab\sin C$

$\quad 23.09 = \dfrac{1}{2}(6.4)(7.9)\sin C$

$\quad 23.09 = 25.28\sin C$

$\quad \sin C = \dfrac{23.09}{25.28}$

$\quad\quad C \approx 66°$

This has a little bit of a twist on the regular area formula. Remember that the angle that is being identified needs to be between the 2 known sides.

Question 15

B $r = 4\%$ p.a.

$\quad = 2\%$ per half-year

$n = 5$ years

$\quad = 10$ half-years

Future value interest factor for 2% at 10 periods

$= 10.9497$

Future value = contribution (C) × future value

interest factor

$\quad 60\,000 = C \times 10.9497$

$\quad\quad\quad C = \dfrac{60\,000}{10.9497}$

$\quad\quad\quad\quad = \$5479.60$

Therefore, Ellie must contribute $5479.60 every 6 months for 5 years to save her $60 000 deposit.

When using the future value interest factor table, always make sure that interest rates and time periods match. Also, remember the formula:

Future value = contribution × future value

interest factor

Finding the contribution is a common twist in these types of questions.

Section II ($\checkmark$ = 1 mark)

Question 16 (1 mark)

Stove $\checkmark$

> The syllabus states 'interpret commonly used symbols and abbreviations on building plans and elevation views'. It's not a Lego piece! Be familiar with symbols on house plans.

Question 17 (2 marks)

sugar : salt : flour

   3 : 1 : 55

3 parts = 42 g

$\therefore$ 1 part = 14 g $\checkmark$

$\therefore$ 55 parts = 770 g $\checkmark$

So 770 g of flour is needed to make the loaf of bread.

> Finding what 1 part is equal to will solve many ratio questions.

Question 18 (1 mark)

Energy label = 386 kWh/year

Vacuum cleaner rating = 2000 W/h

So the vacuum uses 2 kW for every hour of use.

Weekly hours:

$$\frac{386}{2} \div 52$$

$\approx$ 3 hours and 43 minutes per week $\checkmark$

> Remember that the wattage rating of an appliance is the number of watts the device draws after 1 hour of continuous use. This will often need to be converted to kilowatts when working out costs.

Question 19 (2 marks)

$$FV = PV(1 + r)^n$$
$$14\,000 = PV(1 + 0.003)^{60} \checkmark$$

$$PV = \frac{14\,000}{(1.003)^{60}}$$
$$= \$11\,696.93 \checkmark$$

> The words 'compounding' and 'now' are the flags for this question. 'Now' indicates present value is being asked for and 'compounding' is a clue to use the compound interest formula.

Question 20 (2 marks)

2 kg pack: \$2.79/2 kg

$$= \frac{\$2.79}{4}/500\,g = \$0.70/500\,g \checkmark$$

600 g pack: \$0.90/600 g $= \left(\dfrac{0.90}{6} \times 5\right)/500\,g$

$$= \$0.75/500\,g$$

Difference in price per 500 g = \$0.75 − \$0.70

$$= \$0.05 \checkmark$$

> Find common units that can convert to the 500 g (for example, convert to 100 g first, then multiply by 5). This is a routine best buys question. Sometimes there may be 3 choices to consider.

Question 21 (3 marks)

r = 22.1% p.a. = 0.221 p.a.

Daily interest rate = $\dfrac{0.221}{365}$ $\checkmark$

$$FV = PV(1 + r)^n$$
$$= \$3800\left(1 + \frac{0.221}{365}\right)^{26}$$
$$= \$3860.28 \checkmark$$

Interest paid = \$3860.28 − \$3800.00

$$= \$60.28 \checkmark$$

> Always make sure that with credit cards you find the daily rate as a decimal. All percentages need to be divided by 100. It is easy to overlook that the question asks for just the interest paid. Pay attention to these details to avoid losing easy marks.

Question 22 (4 marks)

a   Total = Savings + other expenses

   920 = Savings + 800

   Savings = \$120 $\checkmark$

   Savings as a percentage of income = $\dfrac{120}{920} \times 100\%$

$$\approx 13\% \checkmark$$

b   5% increase in cost of all other items

   = 800 × 1.05

   = \$840 $\checkmark$

   Savings = \$920 − \$840

$$= \$80 \checkmark$$

> Working with percentages is an important skill to master. Budget questions are usually asked with missing items that need to be found.

WORKED SOLUTIONS

Question 23 (2 marks)

$$\text{BAC}_{\text{male}} = \frac{10N - 7.5H}{6.8M}$$

$$0.05 = \frac{10N - 7.5(3)}{6.8(87)}$$

$$29.58 = 10N - 22.5 \checkmark$$

$$10N = 52.08$$

$$N = \frac{52.08}{10}$$

$$= 5.208 \text{ drinks}$$

So Hunter can only drink 5 standard drinks over 3 hours if his BAC is to remain under 0.05. ✓

When the question refers to staying under the limit, make the limit the BAC of the formula and solve. Be prepared to solve any variable within the BAC formula.

Question 24 (2 marks)

Edmonton = −7 and Paris = +5

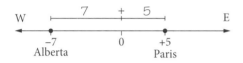

So time difference = 7 + 5

$$= 12 \text{ hours} \checkmark$$

Paris is further east; therefore, it is 12 hours ahead of Edmonton.

Time in Paris when Lani leaves = 9 am + 12 hours

$$= 9 \text{ pm Monday}$$

Time in Paris when Lani arrives
= 9 pm Monday + 14 hours
= 11 am Tuesday ✓

Opposite sides of the prime meridian mean we add the hours. A handy strategy is to always find out the time in the destination city before leaving the city of origin.

Question 25 (4 marks)

a

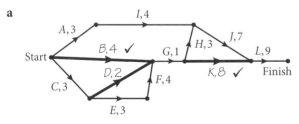

b Forward scanning:

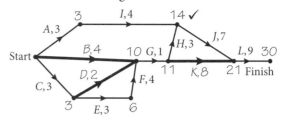

Therefore, the minimum completion time of the project is 30 days. ✓

A forward scan gives the minimum project completion time. Remember that you are looking for the largest number at each vertex as you scan from start to finish.

Question 26 (3 marks)

a Dividend yield $= \dfrac{0.40}{4.30} \times 100\%$

$$\approx 9.3\% \checkmark$$

b Dividend yield is 3.4% higher than last year: 9.3% + 3.4% = 12.7%. Let x = new dividend.

So $12.7\% = \dfrac{x}{4.30}$

$$0.127 = \frac{x}{4.30}$$

$$x \approx \$0.55 \checkmark$$

Increase in dividend = \$0.55 − \$0.40

$$= \$0.15 \text{ per share} \checkmark$$

The formula for dividend yield is not given on the HSC formula sheet. Dividend yield is a common question and is definitely worth practising.

Question 27 (2 marks)

Resting heart rate = 70 beats per minute

High-intensity heart rate = 70 × 2.05

$$= 143.5 \text{ bpm} \checkmark$$

Number of beats for 4 minutes = 143.5 × 4

$$= 574 \text{ beats} \checkmark$$

Heart rate questions come in all forms. This question dealt mainly with percentages. Other questions will work through target heart rate formulas.

Question 28 (3 marks)

First, find the scale of the map:

3.46 cm : 17.3 km

3.46 cm : 1 730 000 cm ✓

346 : 173 000 000

1 : 500 000 ✓

So scaled distance of 24.3 km is:

$$\frac{24.3 \text{ km}}{500\,000} = 0.000\,048\,6 \text{ km}$$
$$= 4.86 \text{ cm} \checkmark$$

> In a ratio, units must be the same and there can be no decimal points, only whole numbers. Conversion between kilometres and centimetres is very important in this question. After the scale is found, the question becomes much easier to answer.

Question 29 (2 marks)

680 km = 430 km (highway) + 250 km (city)

$$\text{Fuel for highway} = \frac{430}{100} \times 8.1$$
$$= 34.83 \text{ L} \checkmark$$

$$\text{Fuel for city} = \frac{250}{100} \times 10.2$$
$$= 25.5 \text{ L}$$

Total fuel used = 38.43 + 25.5
$$= 60.33 \text{ L} \checkmark$$

> It is important to remember that fuel efficiency is measured in L/100 km. This is a standard type of question for fuel usage rates.

Question 30 (4 marks)

a Working through EST and LST for each activity:

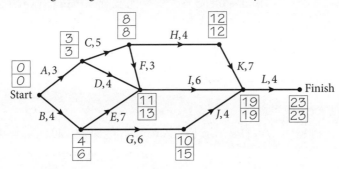

The minimum completion time for the project is 23 days. ✓

b Highlighting the critical path:

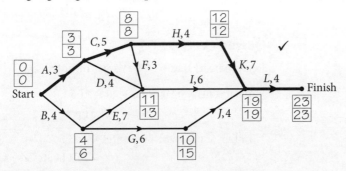

Activities that have 0 float time are those that are on the critical path. Therefore, there are 5 activities that have 0 float time. ✓

c

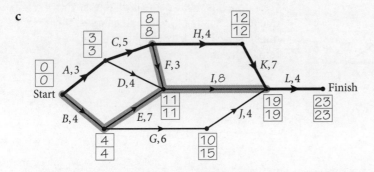

If the duration of activity *I* was changed to 8, it would create 2 more critical paths, meaning 4 more activities would have a float time of 0. ✓

> Activities on the critical path have a float time of 0. Practise finding the critical path quickly and efficiently as this is an important skill in the topic of Networks.

Question 31 (4 marks)

a From graph, when $w = 1$, $A = 5$.

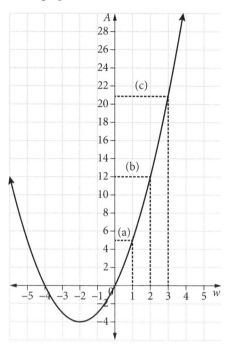

$A = 5\,\text{cm}^2$ when $w = 1\,\text{cm}$ ✓

b From graph, when $w = 2\,\text{cm}$, $A = 12\,\text{cm}^2$.

$A = l \times w$

$12 = l \times 2$

$l = 6\,\text{cm}$ ✓

c From part **b**, when $w = 2\,\text{cm}$, $l = 6\,\text{cm}$.

$P = 2(2) + 2(6)$

$ = 16\,\text{cm}$

From graph, when $w = 3\,\text{cm}$, $A = 21\,\text{cm}^2$.

So $l = 7\,\text{cm}$

$P = 2(3) + 2(7)$

$ = 20\,\text{cm}$ ✓

Difference $= 20 - 16$

$\phantom{\text{Difference}} = 4\,\text{cm}$ ✓

Reading from a non-linear graph is a common HSC question. These questions often include other concepts, such as area and perimeter.

Question 32 (3 marks)

$T = \dfrac{k}{s}$

$9 = \dfrac{k}{80}$

$k = 720$ ✓

It is $720\,\text{km}$ from Town A to Town B.

$T = \dfrac{720}{s}$

$ = \dfrac{720}{90}$

$ = 8\,\text{hours}$ ✓

The car would arrive at 4 pm. ✓

Most inverse variation questions are very similar. All involve finding the value of k and using that value in the formula $y = \dfrac{k}{x}$ to answer the question.

Question 33 (3 marks)

a

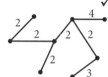 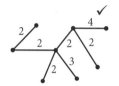

b Length of minimum spanning tree
$= 2 + 2 + 2 + 2 + 2 + 3 + 4 = 17$ ✓

Use Kruskal's or Prim's method. Unless specified, choose the method that is the most comfortable for you and that can be done quickly and accurately.

Question 34 (3 marks)

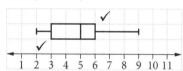

$\text{IQR} = Q_3 - Q_1$

$\phantom{\text{IQR}} = 6 - 3$

$\phantom{\text{IQR}} = 3$ ✓

Straightforward question. Finding the 5-number summary and drawing the corresponding box plot is a Year 11 topic but is usually featured in some form each year in the HSC.

Question 35 (2 marks)

Volume of a prism $= A \times h$

$V = \dfrac{\pi R^2 - \pi r^2}{2} \times 52$

$ = \dfrac{\pi(6)^2 - \pi(4)^2}{2} \times 52$ ✓

$ = 1634\,\text{m}^3$ ✓

Volumes of prisms and cylinders are common questions in the HSC. Usually they will be either composite or half shapes with either a capacity or financial component attached.

Question 36 (3 marks)

First, find the present value of the annuity.
Then, from the compound interest formula,
find the future value of the annuity.

Contribution = $300 per month

r = 3% per month
n = 3 years = 36 months
Present value interest factor = 21.8323 ✓

Present value of annuity
= 300 × 21.8323
= $6549.69 ✓

Future value = $6549.69(1.03)^{36}
            = $18 982.82 ✓

> The present value of an annuity is the single
> amount that can be invested with the same
> interest rate and time period as the future value
> of an annuity with a regular contribution.

Question 37 (3 marks)

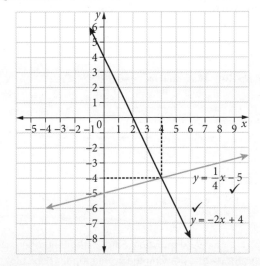

Solution: $x = 4$, $y = -4$ ✓

> Know how to use a table of values or gradient
> and y-intercept to quickly graph straight lines.
> The point of intersection of both lines is the
> solution to the 2 equations.

Question 38 (2 marks)

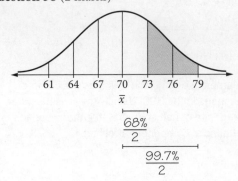

Shaded area = $\dfrac{99.7\%}{2} - \dfrac{68\%}{2}$ (or 13.5% + 2.35%)

= 15.85% ✓

Number of people = 15.85% × 20 000
              = 3170 ✓

> Straightforward normal distribution question.
> It is important to understand how to manipulate
> percentages with the empirical rule. A common
> way to end these questions is to apply the
> percentage to a whole group.

Question 39 (3 marks)

By entering points into the calculator:
$m = 1.282$, $c = 27.859$ ✓

$$f = ms + c$$
$$= 1.282s + 27.859 ✓$$
$$40 = 1.282s + 27.859$$
$$1.282s = 12.141$$
$$s = 9.5 ✓$$

> Be familiar with using the regression tools
> on your calculator. Finding the least-squares
> regression line from given points is an important
> skill to master.

Question 40 (4 marks)

a

	1st pick	2nd pick	Outcomes

```
                      4/12
              5/13   ┌─── G      G G
              ┌── G ─┤
              │      │ 8/12
              │      └─── O      G O  ✓
      ●───────┤
              │      │ 5/12
              │ 8/13 ┌─── G      O G  ✓
              └── O ─┤
                     │ 7/12
                     └─── O      O O  ✓
```

b P(1 lolly of each colour) P(GO) + P(OG)

$$= \frac{5}{13} \times \frac{8}{12} + \frac{8}{13} \times \frac{5}{12}$$

$$= \frac{20}{39} ✓$$

> Because the lollies are not being replaced, the
> totals will change in the 2nd pick. Remember
> when dealing with multiple outcomes to multiply
> along the branches and add the totals of the
> branches together.

9780170459211

Question 41 (3 marks)

a $S = V_0(1 - r)^n$
  $= 28\,950(1 - 0.18)^3$
  $= \$15\,962.10$ ✓

b   Using trial and error:
  $S = 28\,950(1 - 0.18)^{5.3}$
    $= \$10\,112.58$ ✓
  $S = 28\,950(1 - 0.18)^{5.4}$
    $= \$9913.87$

  So it will take 5.4 years to decrease to less than $\$10\,000$. ✓

When finding the value of n, trial and error is used. This is not a common question but it is important to know.

Question 42 (4 marks)

a   Population difference between 4th and 8th years:
  $100(1.05)^8 - 100(1.05)^4 \approx 26$ ✓

b

n	2	7	12	17
P	110	141	180	229	✓

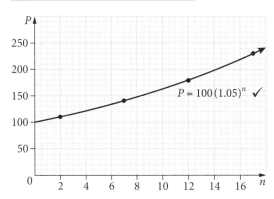

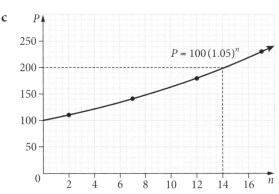

Using the graph, population doubles at approximately 14 years. ✓

Be prepared to draw non-linear graphs and interpolate or extrapolate information from it. It is important to know beforehand what shape each graph should have.

Question 43 (3 marks)

a   Capacity of cut 1 = 110 + 120 + 75
    $= 305$ ✓

b

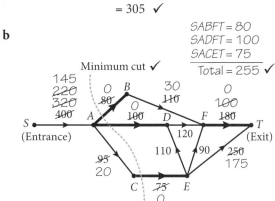

Therefore, the maximum flow is 255.

Routine maximum flow question. When looking for the minimum cut, start by looking to cut through the smallest edges on the network first.

Question 44 (3 marks)

Total original payments:
$\$1422.63 \times 12 \times 25 = \$426\,789$ ✓

Payments for first 3 years:
$\$1422.63 \times 3 \times 12 = \$51\,214.68$

Payments for next 19 years:
$\$1582.63 \times 12 \times 19 = \$360\,839.64$

Total paid after increasing repayment amount after 3 years = $\$412\,054.32$ ✓

Interest saved = $\$426\,789 - \$412\,054.32$
        $= \$14\,734.68$ ✓

Remember that multiplying the monthly repayment by the number of time periods will give the total amount repaid on the loan. This will make finding the interest paid on the loan easy to find.

Question 45 (5 marks)

a

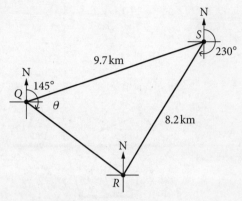

$$\frac{\sin\theta}{8.2} = \frac{\sin 85°}{9.7}$$

$$\sin\theta = \frac{8.2\sin 85°}{9.7}$$

$$\theta = 57° \checkmark$$

Bearing of S from $Q = 145° - 57°$
$$= 088° \checkmark$$

b

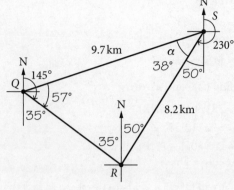

$\alpha = 180° - 85° - 57°$
$\quad = 38° \checkmark$

It is possible to use the sine rule or cosine rule to find QR.

$$\frac{QR}{\sin 38°} = \frac{9.7}{\sin 85°}$$

$$QR = \frac{9.7\sin 38°}{\sin 85°}$$

$$\approx 6.0\,\text{km} \checkmark$$

Total distance $= 6.0 + 8.2 + 9.7$
$$= 23.9\,\text{km} \checkmark$$

This is a harder trigonometry question because it combines bearings and the sine rule. Remember that the point after the word 'from' is the centre of the compass. Also, 2 sides, 2 angles is sine rule and 3 sides, 1 angle is cosine rule.

Mathematics Standard 2

PRACTICE HSC EXAM 2

General instructions	• Reading time: 10 minutes
	• Working time: 2 hours and 30 minutes
	• A reference sheet is provided on page 200 at the back of this book
	• For questions in Section II, show relevant mathematical reasoning and/or calculations

Total marks: 100

Section I – 15 questions, 15 marks

• Attempt Questions 1–15

• Allow about 25 minutes for this section

Section II – 27 questions, 85 marks

• Attempt Questions 16–42

• Allow about 2 hours and 5 minutes for this section

Section I

15 marks
Attempt Questions 1–15
Allow about 25 minutes for this section

Circle the correct answer.

Question 1

Which of the data sets below is closest to the shape of a normal distribution?

A
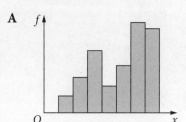

B
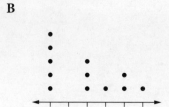

C
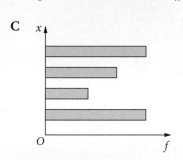

D
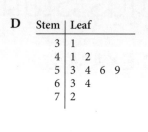

Question 2

A farmer has 510 watermelon plants and 150 strawberry plants.

What is the simplified ratio of strawberry plants to watermelon plants?

A $3:1$

B $5:17$

C $17:5$

D $510:150$

Question 3

Paola earned $488.80 working at the local supermarket. She worked 35 hours normal time and 8 hours at time-and-a-half.

How much is her normal hourly rate of pay?

A $9.75

B $10.40

C $13.97

D $18.34

Question 4

A pen is measured to be 12.1 cm long.

What is the percentage error of the measurement?

A   0.004%

B   0.05%

C   0.1%

D   0.41%

Question 5

Which of the following would most likely have a negative correlation?

A   Amount of eating and body weight

B   Time spent training for a running race and time taken to complete the race

C   Temperature in a city and cold drink sales

D   Number of cars sold and number of basketball game tickets sold

Question 6

Three students are playing basketball, as represented in the diagram below.

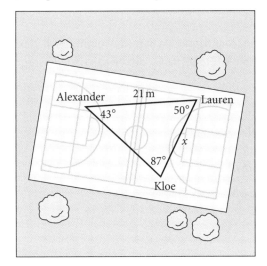

Which formula could be used to calculate how far Lauren is from Kloe?

A $\dfrac{\sin 43°}{x} = \dfrac{21}{\sin 87°}$

B $x^2 = 21^2 + 50^2 - 2 \times 21 \times 50 \times \cos 43°$

C $\dfrac{x}{\sin 43°} = \dfrac{21}{\sin 87°}$

D $\dfrac{x}{\sin 50°} = \dfrac{21}{\sin 43°}$

Question 7

If all graphs use the same scale, which of the following is closest to the correct graph of $y = 12x^2 - 1$?

A
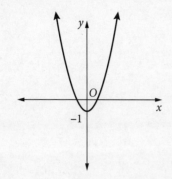

B
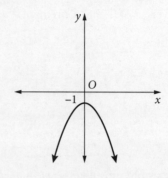

C
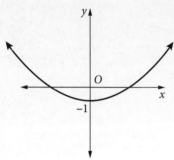

D
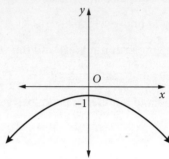

Question 8

Kelly weighs 58 kg and drinks red wine from 6 pm on a Saturday night. Each glass of red wine is equivalent to 1.2 standard drinks. Kelly's blood alcohol content (BAC) can be calculated using the following formula:

$$BAC_{female} = \frac{10N - 7.5H}{5.5M},$$

where N = number of standard drinks
        H = number of hours drinking
        M = female's mass, in kilograms.

If Kelly has had 4 glasses of red wine by 9:30 pm, what is her BAC?

A   0.043

B   0.048

C   0.063

D   0.068

Question 9

Flour is being sold in 3 different-sized packs:

> 2 kg for $4.80
> 1 kg for $2.30
> 250 g for $0.70.

Which pack has the cheapest price per 500 g?

A   250 g pack for $1.20 per 500 g

B   1 kg pack for $1.15 per 500 g

C   2 kg pack for $1.15 per 500 g

D   2 kg pack for $1.20 per 500 g

Question 10

Karl lives in Sydney (UTC +10) and wants to call his mum, who lives in New York (UTC −5). His mum has asked him to call her at 8:30 pm on Tuesday evening New York time.

What is the day and time in Sydney when Karl makes the call?

A   5:30 am Tuesday

B   11:30 am Wednesday

C   8:30 pm Wednesday

D   11:30 pm Wednesday

Question 11

Dots were used to create the following set of patterns:

Figure 1   Figure 2   Figure 3

The relationship is recorded in the table below.

Number of dots in middle row (N)	1	2	3
Number of dots (D)	5	8	11

How many dots are there in the middle row of a pattern with 1568 dots?

A   501

B   520

C   522

D   542

Question 12

Sally's target heart rate when exercising is calculated using the following formula:

$$THR = I \times (MHR − RHR) + RHR,$$

where I = exercise intensity
         MHR = maximum heart rate
         RHR = resting heart rate.

Sally wants to undertake an exercise of 0.85 intensity and achieve a target heart rate of 165.

If she has a resting heart rate of 80 beats per minute, what is her maximum heart rate?

A   170 bpm

B   175 bpm

C   180 bpm

D   185 bpm

Question 13

This grid shows the number of edges connecting 4 vertices of a network.

	A	B	C	D
A	0	0	1	1
B	0	0	1	0
C	1	1	0	1
D	1	0	1	0

Which network diagram is represented by the grid?

A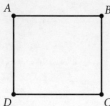

B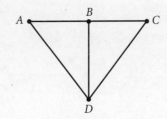

C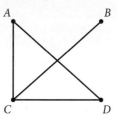

D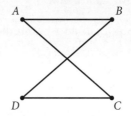

Question 14

In 1986, Monet's painting titled *Meules* sold for $2.53 million. In 2019, the same painting was sold at auction for $110.7 million.

What was the average annual rate of appreciation of the painting, correct to one decimal place?

A   8.3% p.a.

B   12.1% p.a.

C   13.4% p.a

D   16.2% p.a.

Question 15

A landscaper prepares a special mixture of soil and fertiliser in the ratio of 10 : 3. She wants to evenly cover a 52 m² garden bed with the mixture. The layer of soil and fertiliser is to be 10 cm thick.

If 1 m³ of the mixture weighs 1000 kg, then how many kilograms of fertiliser will she need?

A   922 kg

B   1100 kg

C   1200 kg

D   1334 kg

Section II

85 marks
Attempt Questions 16–42
Allow about 2 hours and 5 minutes for this section

- Answer the questions in the spaces provided. These spaces provide guidance for the expected length of response.
- Your responses should include relevant mathematical reasoning and/or calculations.

Question 16 (1 mark)

Clark's formula is used to determine the dosage amount of medicine for children. The formula is given by:

$$\text{Child dosage} = \frac{\text{weight of child (kg)} \times \text{adult dosage (mg)}}{70}$$

If the adult dosage is 250 mg and the child weighs 28 kg, what is the recommended child dosage?   1 mark

Question 17 (2 marks)

Xan gets a job at his local supermarket. His normal rate of pay is $14.50 per hour. This week he was paid for working 12 hours normal time and 5 hours time-and-a-half. The local butcher asked Xan to work for him instead and told him that he could earn exactly what he earned this week just by working 11 hours.

What rate of pay is the butcher willing to pay Xan per hour?   2 marks

Question 18 (3 marks)

The network below shows the activities that need to occur for a project to be completed.

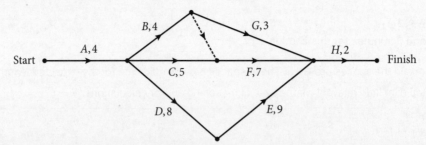

Complete the activity table below for the network shown.   3 marks

Activity	Duration	Immediate predecessor
A	4	
	4	A
C		
D		A
	9	
F		
G		
H	2	

Question 19 (2 marks)

A rectangular piece of wood is measured as having a length of 14.5 cm and a width of 32.7 cm.

Taking into account the limits of accuracy of both length and width measurements, what is the upper limit of the actual area of the chopping board, correct to two decimal places?   2 marks

Question 20 (2 marks)

Caleb has a bag of green and blue marbles. There are 40 green marbles in the bag. There are 40% more blue marbles in the bag than green marbles.

What is the simplified ratio of blue to green marbles?   2 marks

Question 21 (3 marks)

A set of data has a lower quartile of 40 and an upper quartile of 54.

What is the maximum possible range if there are no outliers in the data set?   3 marks

Question 22 (3 marks)

Company A shares have a market price of $8.80 and pay a dividend of $0.96 per share. Company B shares have a market price of $3.90 and pay a dividend of $0.72 per share.

Show by calculations which shares are a better buy from a dividend perspective.   3 marks

Question 23 (3 marks)

The scores for an English test and a History test are shown in the parallel box plot.

Compare and contrast the two data sets, referring to skewness, measures of central tendency and spread.   3 marks

_____

_____

_____

_____

_____

_____

_____

Question 24 (3 marks)

Padang, Indonesia is located at $(0°, 100°\,E)$ and Quito, Ecuador is located at $(0°, 78°\,W)$. Both cities lie on the Equator, which is a great circle with radius 6400 km. A plane takes 22 hours to fly non-stop from Padang to Quito.

Calculate the difference in longitude between the cities, and hence the plane's average speed to the nearest km/h.   3 marks

_____

_____

_____

_____

_____

_____

_____

_____

Question 25 (3 marks)

Calculate the area of the shaded region in the sector below, correct to two decimal places.   3 marks

Question 26 (2 marks)

In a normally distributed set of scores, the mean is 35 and the standard deviation is 3.

What percentage of scores lie between 26 and 29?   2 marks

Question 27 (3 marks)

At the local showground, 12 stalls need to be connected to electricity. Each edge represents the length of cabling, in metres, needed to connect each stall. All of the sites are to be connected with the smallest length of electrical cable possible.

a   On the diagram, draw or highlight where the cables need to be installed so that all stalls are connected with the smallest length of cable.   2 marks

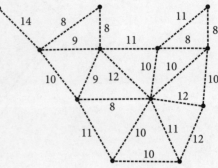

b   If the cost to install the cable is $30 per linear metre, how much will this cost the showground organisers?   1 mark

_____

_____

_____

Question 28 (4 marks)

The table shows the required monthly payment amount for a loan of $100 at various interest rates and loan terms.

Monthly payment amount by loan term (months) and interest rate (%)

	12	24	36	48	60	72	84
2%	$8.42	$4.25	$2.86	$2.17	$1.75	$1.48	$1.28
4%	$8.51	$4.34	$2.95	$2.26	$1.84	$1.56	$1.37
6%	$8.61	$4.43	$3.04	$2.35	$1.93	$1.66	$1.46
8%	$8.70	$4.52	$3.13	$2.44	$2.03	$1.75	$1.56
10%	$8.79	$4.61	$3.23	$2.54	$2.12	$1.85	$1.66
12%	$8.88	$4.71	$3.32	$2.63	$2.22	$1.96	$1.77
14%	$8.98	$4.80	$3.42	$2.73	$2.33	$2.06	$1.87
16%	$9.07	$4.90	$3.52	$2.83	$2.43	$2.17	$1.99

A loan of $70 000 is taken out at 6% p.a. interest.

a   What is the difference in monthly repayments between 3- and 5-year terms?   2 marks

_____

_____

_____

_____

_____

_____

_____

Question 28 continues on page 167

9780170459211

Question 28 (continued)

b How much total interest is saved by taking the 3-year term repayments rather than the 5-year term repayments?

2 marks

Question 29 (4 marks)

The network below shows all of the activities that are needed to complete a project. Each activity is 3 weeks in duration.

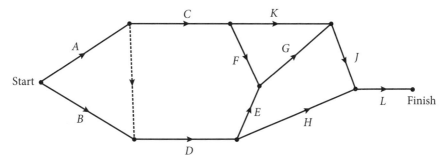

a If activity *K* is reduced by 1 week, calculate the effect on the project's completion time.

2 marks

b What would be the effect on the project's minimum completion time if the durations of all of the activities were reduced by 1 week?

2 marks

Question 30 (5 marks)

The population of Krusetown was recorded every 2 years from 2000 to 2022. The data is shown in the table below.

Year	2000	2002	2004	2006	2008	2010	2012	2014	2016	2018	2020	2022
Population (thousands)	14	14.5	16	18.5	19.5	21	23	24	25	25.5	27	28

a   Plot the points in the table on the grid below.   2 marks

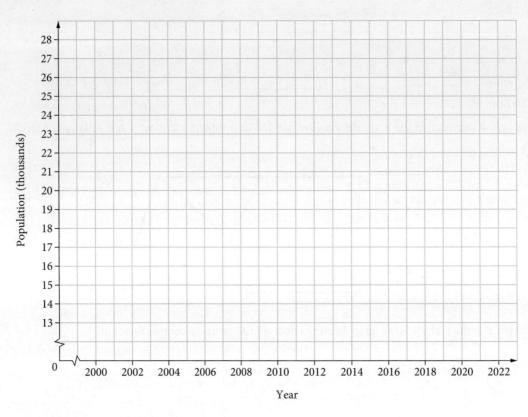

b   On the grid above, draw a line of best fit by eye.   1 mark

c   What does the gradient mean in this context?   1 mark

_____

_____

_____

d   Calculate Pearson's correlation coefficient for the data, correct to four decimal places.   1 mark

_____

_____

Questions 16–30 are worth 43 marks in total (Section II halfway point)

Question 31 (3 marks)

Jala performed a radial survey of the triangular piece of land shown.

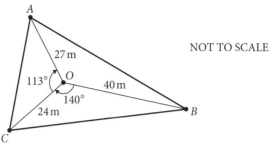

NOT TO SCALE

a   Find the area of ΔAOB, correct to one decimal place.   2 marks

b   If ΔAOB makes up 46% of the total area of the land, what is the total area of ΔABC?   1 mark

Question 32 (2 marks)

Eleazar takes a Maths and an English test. Both class test results are normally distributed.
In Maths he works out that his z-score is 1.8. In English he scored a mark of 87.
The mean for the English test was 77.5 and the standard deviation was 5.

In which test did Eleazar perform relatively better when compared to the rest of the class?   2 marks

Question 33 (5 marks)

Natalia buys a brand new computer for $5000 for her photography business. She uses the declining-balance method to calculate the amount of depreciation of the computer. She claims this amount as a tax deduction each year. She can only claim the depreciation for that particular year.

a How much does she claim in the 2nd year if the rate of depreciation is 27% p.a.?   2 marks
Round to the nearest dollar.

b Natalia earned $90 000 for the last financial year. The depreciation amount for her computer in part **a** is the only tax deduction she claims.

Using the tax table below and ignoring the Medicare levy, how much tax does she need to pay?   3 marks

Income tax rates for Australian residents

Taxable income	Tax on this income
$0 – $18 200	Nil
$18 201 – $37 000	19c for each $1 over $18 200
$37 001 – $87 000	$3572 plus 32.5c for each $1 over $37 000
$87 001 – $180 000	$19 822 plus 37c for each $1 over $87 000
$180 001 and over	$54 232 plus 45c for each $1 over $180 000

Question 34 (4 marks)

Levron Games received a credit card statement on 2 November. There is a no interest-free period and interest is compounded daily. (Note: There are 31 days in October.)

Tri-Bank
"We're not Pointless"

STATEMENT OF ACCOUNT

Account Number: 122-233-344-455
Statement Date: 02/11/2020
Period Covered: 01/09/2020 to 01/11/2020

Page 1 of 1

Levron Games
23 Champion Road
Lakewood NSW
2020

Camden Branch

Opening Balance:	$0.00
Total Credit Amount:	$1450.00
Annual Percentage Rate:	18.1% p.a.
Closing Balance:	$1450.00 + Interest
Minimum Payment:	$50 or 5% of amount owing (whichever is greater)

Transactions

Date	Description	Credit	Debit	Balance
12/10/2020	100 Bottles of Fruity Fizz	$1450.00		$1450.00
02/11/2020	Interest Charged	?		?

---End of transactions---

Note: Interest is charged on amounts from and including the date of purchase up to and including the date of payment.

a If the statement was paid in full on 2 November, what was the total amount paid?   2 marks

b What was the minimum payment required on this account?   2 marks

Question 35 (2 marks)

The following network shows Town A and Town K connected by a series of roads. The edge weights represent distances, in kilometres.

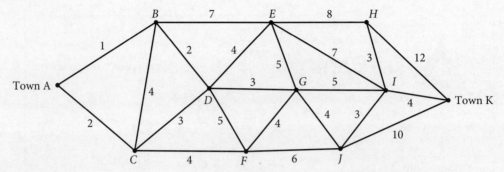

What is the shortest distance between Town A and Town K? List the vertices of the shortest path from Town A to Town K.

2 marks

_____

_____

_____

Question 36 (1 mark)

A deck of 52 playing cards contains 4 kings. 3 cards are drawn at random and placed on the table after each draw.

What is the probability that all 3 cards drawn are kings?

1 mark

_____

_____

_____

_____

_____

_____

Question 37 (4 marks)

The network below shows the capacity of each edge.

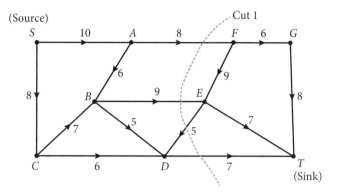

a What is the capacity of cut 1?   1 mark

b What is the maximum flow capacity of the network?   2 marks

c What is the smallest number of edges that need to be increased by 2 to increase the maximum flow by 4?   1 mark

Question 38 (4 marks)

Joel is buying his first car. He takes out a loan for $5000 and agrees to repay the loan back in monthly repayments at 12% p.a. for 4 years.

Use the present value of an annuity table to determine how much interest he will pay on the loan.   4 marks

Present value of an annuity of $1 per period for *n* periods

Period	1%	2%	3%	4%	5%	6%	7%	8%	9%	10%
30	25.8077	22.3965	19.6004	17.2920	15.3725	13.7648	12.4090	11.2578	10.2737	9.4269
31	26.5423	22.9377	20.0004	17.5885	15.5928	13.9291	12.5318	11.3498	10.3428	9.4790
32	27.2696	23.4683	20.3888	17.8736	15.8027	14.0840	12.6466	11.4350	10.4062	9.5264
33	27.9897	23.9886	20.7658	18.1476	16.0025	14.2302	12.7538	11.5139	10.4644	9.5694
34	28.7027	24.4986	21.1318	18.4112	16.1929	14.3681	12.8540	11.5869	10.5178	9.6086
35	29.4086	24.9986	21.4872	18.6646	16.3742	14.4982	12.9477	11.6546	10.5668	9.6442
36	30.1075	25.4888	21.8323	18.9083	16.5469	14.6210	13.0352	11.7172	10.6118	9.6765
37	30.7995	25.9695	22.1672	19.1426	16.7113	14.7368	13.1170	11.7752	10.6530	9.7059
38	31.4847	26.4406	22.4925	19.3679	16.8679	14.8460	13.1935	11.8289	10.6908	9.7327
39	32.1630	26.9026	22.8082	19.5845	17.0170	14.9491	13.2649	11.8786	10.7255	9.7570
40	32.8347	27.3555	23.1148	19.7928	17.1591	15.0463	13.3317	11.9246	10.7574	9.7791
41	33.4997	27.7995	23.4124	19.9931	17.2944	15.1380	13.3941	11.9672	10.7866	9.7991
42	34.1581	28.2348	23.7014	20.1856	17.4232	15.2245	13.4524	12.0067	10.8134	9.8174
43	34.8100	28.6616	23.9819	20.3708	17.5459	15.3062	13.5070	12.0432	10.8380	9.8340
44	35.4555	29.0800	24.2543	20.5488	17.6628	15.3832	13.5579	12.0771	10.8605	9.8491
45	36.0945	29.4902	24.5187	20.7200	17.7741	15.4558	13.6055	12.1084	10.8812	9.8628
46	36.7272	29.8923	24.7754	20.8847	17.8801	15.5244	13.6500	12.1374	10.9002	9.8753
47	37.3537	30.2866	25.0247	21.0429	17.9810	15.5890	13.6916	12.1643	10.9176	9.8866
48	37.9740	30.6731	25.2667	21.1951	18.0772	15.6500	13.7305	12.1891	10.9336	9.8969
49	38.5881	31.0521	25.5017	21.3415	18.1687	15.7076	13.7668	12.2122	10.9482	9.9063
50	39.1961	31.4236	25.7298	21.4822	18.2559	15.7619	13.8007	12.2335	10.9617	9.9148

Question 39 (4 marks)

Katelyn decides to set up a furniture business to sell bedside tables. Below are the income (*I*) and cost (*C*) functions for her business.

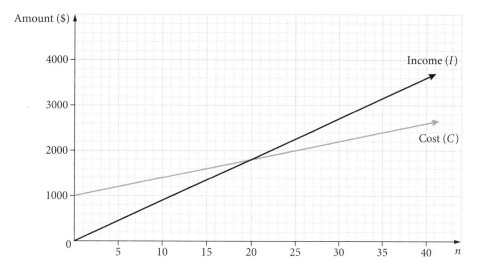

a   Write the cost function in the form $C = mn + c$ for the cost of making *n* tables.   2 marks

b   For how much does Katelyn plan to sell each bedside table?   1 mark

c   If Katelyn sells 30 bedside tables, how much profit or loss will she incur?   1 mark

Question 40 (5 marks)

Perla wants to take a holiday and leave from Baan to visit Mamaks. Unfortunately, there is no direct flight, so she is required to fly to Ali first.

She flies to Ali from Baan on a bearing of 120°. The plane flies at an average speed of 800 km/h for 5 hours.

From Ali she then flies to Mamaks on a bearing of 040°. The plane flies for 7.5 hours at an average speed of 950 km/h.

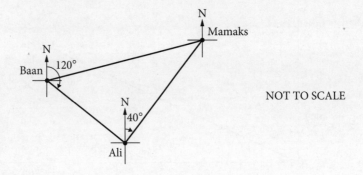

NOT TO SCALE

If a plane flew directly from Baan to Mamaks at an average speed of 800 km/h, how long would the flight be, correct to the nearest minute?   5 marks

Question 41 (4 marks)

A movie theatre can seat 300 people. Tickets currently cost $20. Some research found that for every $1 increase in price, 10 fewer people are expected to attend. The graph below shows the expected revenue for every increase by x in ticket prices.

For example, when $x = 8$, the price increase is $8, so a movie ticket would cost $28 and the revenue looks to be around $6160.

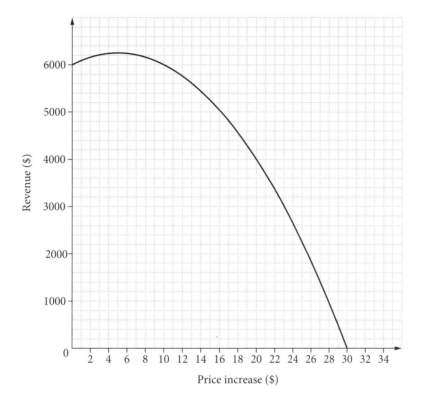

a   What price should the theatre charge for tickets to maximise revenue?   1 mark

b   Why does this graph not apply for values of x greater than 30?   1 mark

c   What would be the difference in revenue if tickets were sold for $36 rather than $40?   2 marks

Question 42 (4 marks)

A company buys a new warehouse to improve their manufacturing capacity. Their plan is to save money by collecting rainwater from the entire area of the rectangular roof. The measurements indicated on the image are map measurements.

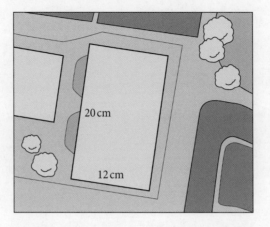

20 cm

12 cm

NOT TO SCALE

a   If 20 cm on the image represented 200 m, what would be the scale for the diagram?   1 mark

b   If the company's goal is to collect 1800 kL of water, how much rain (in mm) must fall?   3 marks

END OF PAPER

WORKED SOLUTIONS

WORKED SOLUTIONS

Section I (1 mark each)

Question 1

D

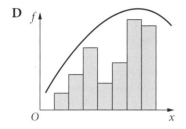

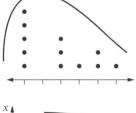

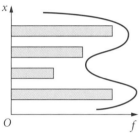

Stem	Leaf
3	1
4	1 2
5	3 4 6 9
6	3 4
7	2

Drawing a line around the peaks of the data shows which set has the shape of a normal distribution. Other questions may refer to skewness or symmetry.

Question 2

B strawberry : watermelon
$$150 : 510$$
$$5 : 17$$

This is a simple ratio question. Remember that order is always important.

Question 3

B Let x be the normal hourly rate of pay.

$$\$488.80 = 35x + 8 \times x \times 1.5$$
$$\$488.80 = 47x$$
$$x = \frac{488.80}{47}$$
$$= \$10.40$$

Work out how many hours in total the employee is paid taking into account any time-and-a-half and double-time hours.

Question 4

D Precision = 0.1
(because the measurement has one decimal place)

Absolute error = 0.05   (half of the precision)

$$\% \text{ Error} = \frac{0.05}{12.1} \times 100\%$$
$$= 0.41\%$$

Percentage error is a common HSC question.

Question 5

B

Negative correlation refers to variables moving in the opposite direction. Theoretically, the more time spent training, the less time it will take the athlete to complete the race. It is more likely to have a negative correlation.

Question 6

C Sine rule refers to 2 sides, 2 angles. In this case, the angle opposite the side that is required is 43°. 21 is opposite the angle 87°.

Sine and cosine rule problems are common questions in the HSC.

Question 7

A The 12 in front of x^2 signifies a narrower parabola, and because 12 is positive the parabola is concave up.

Question 8

D $N = 4 \times 1.2$

$= 4.8$ standard drinks

$BAC_{female} = \dfrac{10(4.8) - 7.5(3.5)}{5.5(58)}$

$= 0.068$

Straightforward BAC substitution question.

Question 9

B $2\,kg: \dfrac{4.80}{4} = \$1.20$ per $500\,g$

$1\,kg: \dfrac{2.30}{2} = \$1.15$ per $500\,g$

$250\,g: 0.70 \times 2 = \$1.40$ per $500\,g$

So $1\,kg$ size is the cheapest price per $500\,g$.

Straightforward best buys question. Practise your unit pricing skills.

Question 10

B Time difference = 15 hours

Karl is further east, so his time is ahead of New York time.

8:30 pm Tuesday (NY) + 15 h

= 11:30 am Wednesday (SYD)

Any city further east will be ahead in terms of time. Always work out what the time difference is and who is further east before starting these types of questions.

Question 11

C Any linear pattern will follow $y = mx + c$.

$m = \dfrac{8 - 5}{2 - 1}$

$= 3$

So $D = 3N + c$

When $N = 1$, $D = 5$.

$5 = 3(1) + c$

$c = 2$

So $D = 3N + 2$

$1568 = 3N + 2$

$N = \dfrac{1566}{3}$

$= 522$ dots in middle row

Establishing rules from linear patterns is a common question. Make sure you can find a rule from a table.

Question 12

C $THR = I \times (MHR - RHR) + RHR$

$165 = 0.85(MHR - 80) + 80$

$85 = 0.85(MHR - 80)$

$MHR - 80 = \dfrac{85}{0.85}$

$MHR - 80 = 100$

$MHR = 180$ bpm

Heart rate questions are usually algebraic substitution questions. Make sure to read all parts of the question carefully.

Question 13

C

0s represent no connection. Remember that edges can cross each other if necessary.

Question 14

B $FV = PV(1 + r)^n$

$110\,700\,000 = 2\,530\,000(1 + r)^{33}$ $(2019 - 1986 = 33)$

$(1 + r)^{33} = \dfrac{110\,700\,000}{2\,530\,000}$

$1 + r = \sqrt[33]{\dfrac{110\,700\,000}{2\,530\,000}}$

$r = 1.1213\ldots - 1$

$= 0.1213\ldots$

$\approx 12.1\%$

Appreciation works exactly the same way as compound interest. Remember that if you want to undo a power, you need to take the root of that power.

Question 15

C $V = Ah$

$= 52 \times 0.1$

$= 5.2\,m^3$

$5.2\,m^3 = 5200\,kg$ of mixture

Total parts $= 10 + 3$

$= 13$

$1\ part = \dfrac{5200}{13}$

$= 400\,kg$

So fertiliser: 3 parts = $1200\,kg$.

An even covering of the mixture with a uniform thickness is the volume of a prism. Make sure the units are the same (for example, converting centimetres to metres).

WORKED SOLUTIONS

Section II (✓ = 1 mark)

Question 16 (1 mark)

Child dosage

$= \dfrac{\text{weight of child (kg)} \times \text{adult dosage (mg)}}{70}$

$= \dfrac{28 \times 250}{70}$

$= 100\,\text{mg}$ ✓

Straightforward algebraic substitution question.

Question 17 (2 marks)

Week's pay $= 12 \times \$14.50 + 5 \times \$14.50 \times 1.5$

$\qquad\qquad = \$282.75$ ✓

Butcher's pay rate $= \dfrac{\$282.75}{11}$

$\qquad\qquad\qquad = \$25.70 \text{ per hour}$ ✓

Always try to set out working in a logical and systematic manner. Look out for time-and-a-half rates and penalty rates in general.

Question 18 (3 marks)

Activity	Duration	Immediate predecessor(s)	
A	4	–	✓
B	4	A	
C	5	A	
D	8	A	
E	9	D	✓
F	7	B, C	
G	3	B	
H	2	E, F, G	✓

Dummy activities are still part of the activity preceding it. In this case the dummy activity is still activity B.

Question 19 (2 marks)

Precision = 0.1 cm

Absolute error = 0.05 cm ✓

Upper limit of area $= 14.55 \times 32.75$

$\qquad\qquad\qquad\quad = 476.51\,\text{cm}^2$ ✓

Questions about limits of accuracy always start with identifying the precision of the measurement and then adding or subtracting half of that precision to or from the measurement taken.

Question 20 (2 marks)

40 green marbles

40% more blue $= 40 \times 1.4$

Blue marbles = 56 ✓

blue : green $= 56 : 40$

$\qquad\qquad = 7 : 5$ ✓

Question 21 (3 marks)

IQR $= 54 - 40$

$\qquad = 14$ ✓

Lower outlier boundary $= Q_1 - 1.5 \times \text{IQR}$

$\qquad\qquad\qquad\qquad = 40 - 1.5 \times 14$

$\qquad\qquad\qquad\qquad = 19$

Upper outlier boundary $= Q_3 + 1.5 \times \text{IQR}$

$\qquad\qquad\qquad\qquad = 54 + 1.5 \times 14$

$\qquad\qquad\qquad\qquad = 75$ ✓

Maximum range $= 75 - 19$

$\qquad\qquad\qquad = 56$ ✓

The formula for outliers is important to know, and also found on the HSC reference sheet. These boundaries represent the smallest and largest values in a data set that would not be considered outliers.

Question 22 (3 marks)

Company A dividend yield $= \dfrac{0.96}{8.80} \times 100$

$\qquad\qquad\qquad\qquad = 10.9\%$ ✓

Company B dividend yield $= \dfrac{0.72}{3.90} \times 100$

$\qquad\qquad\qquad\qquad = 18.5\%$ ✓

Therefore, Company B has a higher dividend yield, meaning these shares are a better buy from a dividend perspective. ✓

Dividend yield is a formula that is usually not given on the HSC reference sheet. Be familiar with the concept because questions related to it are asked often.

Question 23 (3 marks)

	English	**History**	
Skewness	Slightly positively skewed	Negatively skewed	✓
Median	30	50	
IQR	60 − 20 = 40	60 − 40 = 20	

The median in History was higher than that in English, so the History scores were better than the English scores. ✓

The IQR in English was higher than that in History, so the English scores were more spread out. ✓

> When comparing data, it is often helpful to use a table. It is important to use comparison words such as 'higher' or 'lower' as the question often asks to compare or contrast. For 3 marks, you need to mention 3 points, but you should avoid writing long-winded paragraphs. Keep things brief and to the point.

Question 24 (3 marks)

Difference in longitude = 100° + 78°

$$= 178° \checkmark$$

$$\text{Distance} = \frac{\theta}{360°} \times 2 \times \pi \times r$$

$$= \frac{178°}{360°} \times 2 \times \pi \times 6400$$

$$= 19\,882.7908\ldots \text{km} \checkmark$$

$$S = \frac{D}{T}$$

$$= \frac{19\,882.7908\ldots}{22}$$

$$= 904 \text{ km/h} \checkmark$$

> Finding distances along a great circle can be done using the arc length formula. This means that the 2 locations will always lie on the equator or on the same line of longitude.

Question 25 (3 marks)

Shaded area = area of sector − area of triangle

$$= \frac{40°}{360°} \times \pi \times 9^2 - \frac{1}{2} \times 3 \times 3 \times \sin 40°$$
✓ ✓

$$= 25.38 \text{ m}^2 \checkmark$$

> To find the area of a shaded part, always find the area of the bigger shape and subtract the area of the smaller shape.

Question 26 (2 marks)

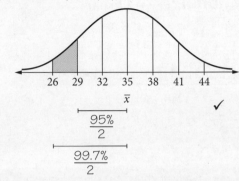

Shaded area $= \frac{99.7}{2} - \frac{95}{2}$

$$= 49.85 - 47.5$$

$$= 2.35\%$$

So percentage of scores between 26 and 29 = 2.35% ✓

> Common type of question. Knowing how to manipulate the percentages on a normal distribution is a very important skill. Practise finding different combinations of each section.

Question 27 (3 marks)

a

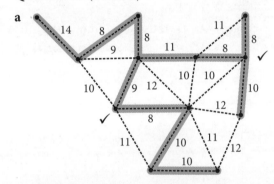

b Minimum length = 100 m

So cost = 100 × 30

$$= \$3000 \checkmark$$

> When finding the minimum spanning tree, start by finding the smallest weights and cross off one at a time, making sure that no cycles are formed.

Question 28 (4 marks)

a 3 years at 6% = 700 × $3.04
                = $2128.00 ✓

   5 years at 6% = 700 × $1.93
                = $1351

   Difference = $2128 − $1351
             = $777 ✓

b 3-year term = $2128 × 12 × 3
   3-year total payments = $76 608
   3-year total interest = $6608 ✓

   5-year term = $1351 × 12 × 5
   5-year total payments = $81 060
   5-year total interest = $11 060

   Interest saving for 3-year term
   = $11 060 − $6608
   = $4452 ✓

   In financial maths problems, you need to be good at understanding the logic behind the 'finance' as well as the 'maths'. Make sure to read all of the information in the table, in this case the loan amount the repayments are referring to. Remember that the monthly repayment can be used to find out the total amount paid for the loan.

Question 29 (4 marks)

a To find out which activities will affect the project's completion time, find the critical path.

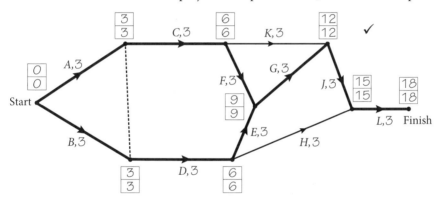

Activity K is not on the critical path, meaning there will be no effect on the project's completion time with a reduction in activity K. ✓

b

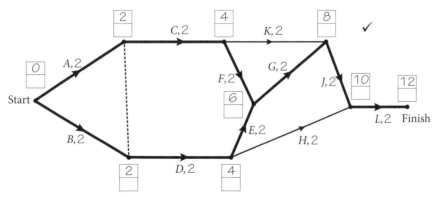

Reducing all the activities to 2 will result in a minimum completion time of 12, which is a reduction of 6 weeks. ✓

A forward scan will reveal the minimum completion time of a project. Remember that a reduction of any activity that is not on the critical path means that it will have no effect on the project's minimum completion time.

Question 30 (5 marks)

a and **b**

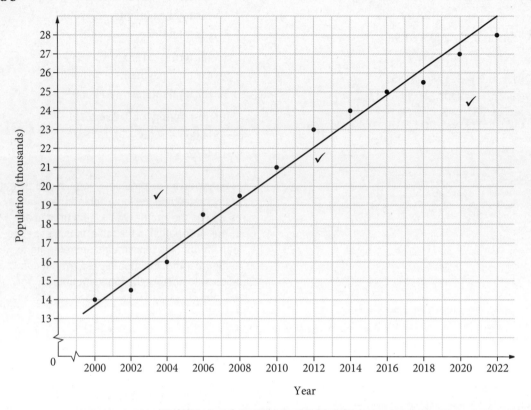

c   The gradient is the average annual rate of increase in the population of Krusetown. ✓

d $r = 0.9924$ ✓

> When drawing a line of best fit by eye, look for a balance of points above and below the line as well as trying to draw the line through as many of the points as possible. Explaining the meaning of the gradient in context is a common exam question.

Question 31 (3 marks)

a $\angle AOB = 360° - 113° - 140°$

$= 107°$ ✓

$A = \dfrac{1}{2}ab\sin C$

$= \dfrac{1}{2} \times 27 \times 40 \times \sin 107°$

$= 516.4\,\text{m}^2$ ✓

b $46\% = 516.4\,\text{m}^2$

$1\% = 11.226\,\text{m}^2$

$100\% = 1122.6\,\text{m}^2$ ✓

> Practise applying the area of a triangle sine formula.

Question 32 (2 marks)

Maths z-score = 1.8

English z-score = $\dfrac{87 - 77.5}{5}$

$= 1.9$ ✓

Therefore, Phillip performed better in English as his z-score was higher than his Maths z-score. ✓

> Comparing performance using z-scores is a common question. Know how to use the z-score formula and make sure you make a concluding statement that answers the question directly.

Question 33 (5 marks)

a　1st year: $S = 5000(1 - 0.27)^1$

Salvage value = $3650

2nd year: $S = 5000(1 - 0.27)^2$ or $3650(1 - 0.27)^1$

Salvage value = $2664.50 ✓

Depreciation amount claimed in the 2nd year

= $3650 − $2664.50

= $985.50

≈ $986 ✓

b　Taxable income = $90 000 − $986

= $89 014 ✓

Tax bracket: $87 001 − $180 000

Tax payable

= $19 822 + 0.37 × ($89 014 − $87 000) ✓

= $20 567.18 ✓

Tax tables are a common HSC question. There are many variations, which may include Medicare levy or PAYG tax. Be familiar with all of these and know how to use the tax table correctly.

Question 34 (4 marks)

a　Days = 31 − 12 + 1 + 2

= 22 ✓

$$A = \$1450\left(1 + \frac{0.181}{365}\right)^{22}$$

Total amount to pay = $1465.90 ✓

Credit card statements usually give you the dates of purchase and payment; however, it is important to know how to count the number of days of interest. Including *both* dates means '+1' when totalling days. Make sure that the interest rate is a daily rate when using it in the compound interest formula.

b　Minimum payment

= $50 or 5% of amount owing
　(whichever is greater)

5% × $1465.90 = $73.30 ✓

This is greater than $50, therefore minimum payment is $73.30. ✓

Question 35 (2 marks)

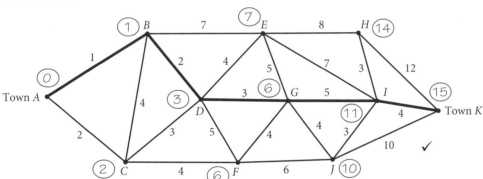

The shortest distance from Town *A* to Town *K* is 15 km. The shortest path is *ABDGIK*. ✓

When determining the shortest path, find the shortest distance to each vertex first (as shown in the diagram). When the end is reached, work backwards and highlight the path.

Question 36 (1 mark)

Probability of 3 kings in a row:

$$P(KKK) = \frac{4}{52} \times \frac{3}{51} \times \frac{2}{50} = \frac{1}{5525} \quad ✓$$

This question involving 3-step events could be represented by the branches on a tree diagram. When calculating probabilities of multi-stage events, multiply the probabilities along the branches. This is a non-replacement question, so the total sample space changed after each event because the cards were not replaced.

Question 37 (4 marks)

a   Capacity of cut 1 = 8 + 9 + 7

$\qquad\qquad\qquad\quad$ = 24 ✓

b   By minimum cut or reducing edges:

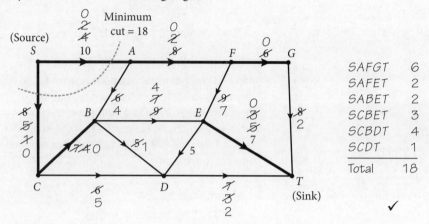

SAFGT	6
SAFET	2
SABET	2
SCBET	3
SCBDT	4
SCDT	1
Total	**18**

✓

The maximum flow capacity of the network is 18. ✓

c

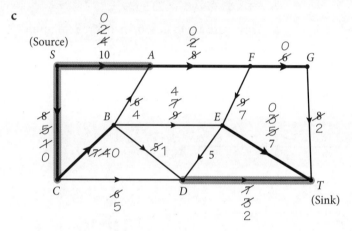

Edges with '0' on them are edges already at maximum capacity. To increase the flow by 4, the grey *SA*, *SC* and *DT* edges need to be increased by 2. Therefore, the minimum number of edges that need to increase by 2 to increase the maximum flow by 4 is 3. ✓

> Finding the maximum capacity through reducing edges is an important skill as the minimum cut will not help you with a question such as part **c**. Highlighting those edges that reach maximum capacity after going through the process will quickly show which edges need to increase capacity to increase maximum flow.

Question 38 (4 marks)

r = 12% p.a. = 1% per month

n = 4 years = 48 time periods

Present value interest factor = 37.9740 ✓

Present value = contribution (P) × present value
$\qquad\qquad\qquad\qquad\quad$ interest factor

$5000 = P \times 37.9740$

$\qquad P$ = $131.67 per month ✓

Total payments = $131.67 × 48

$\qquad\qquad\qquad\quad$ = $6320.16 ✓

So interest = $6320.16 − $5000

$\qquad\qquad\quad$ = $1320.16 ✓

> The present value of an annuity is commonly used in the context of loans. The monthly repayment of a loan is the contribution to an annuity. The loan amount that is given upfront is the present value. Therefore, finding the monthly repayment is as simple as dividing the loan amount by the present value interest factor.

Question 39 (4 marks)

a

$$m = \frac{\text{rise}}{\text{run}}$$

$$= \frac{800}{20}$$

$$= 40 \checkmark$$

$c = 1000$ (where the cost function intersects
the $ axis)

So $C = 40n + 1000$ ✓

b

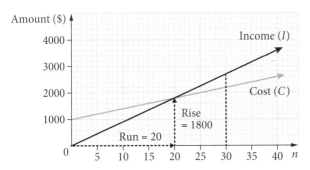

Looking at gradient of income function:

$$m = \frac{1800}{20} = \$90$$

So bedside tables will be sold for $90 each. ✓

c Looking at the graph or formulas at $n = 30$,
the profit will be the difference between
income and cost.

Profit = $2700 – $2200

   = $500 ✓

Cost and income function questions are
commonly asked. Know the definition of break-
even point as well as where the profit and loss
sections of the graph are.

Question 40 (5 marks)

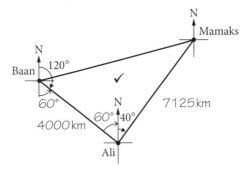

Baan to Ali = 800 km/h × 5 hours

   = 4000 km ✓

Ali to Mamaks = 950 km/h × 7.5 hours

   = 7125 km ✓

$$c^2 = a^2 + b^2 - 2ab\cos C$$

$$BM^2 = 4000^2 + 7125^2 - 2 \times 4000 \times 7125 \times \cos 100°$$

$$BM = \sqrt{76\,663\,571.13}$$

   $$= 8755.77\,\text{km} \checkmark$$

If plane is travelling at 800 km/h:

$$T = \frac{D}{S}$$

$$= \frac{8755.77\ldots}{800}$$

$$\approx 10 \text{ hours and 57 minutes} \checkmark$$

Questions that combine bearings with cosine
or sine rules represent the harder trigonometry
questions in the course. The added twist in this
question is working with speed. Make sure to set
your work out in a logical manner.

Question 41 (4 marks)

a

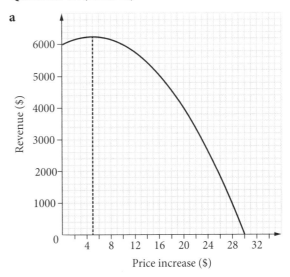

Maximum occurs when $x = 5$.

So ticket charge to maximise revenue

= $20 + $5

= $25 ✓

b If x is greater than 30, this means a price increase of at least $30, so $20 + $30 = $50. It also means that the revenue will be $0 or negative, which is impractical and why the graph doesn't apply for values greater than 30. ✓

c At $36 ($x = 16$), revenue = $5040 ✓

At $40 ($x = 20$), revenue = $4000

Difference in revenue = $1040 ✓

> Questions requiring reading from a quadratic model in a particular context have appeared in numerous HSC papers. Be prepared to identify parts of the graph that are not applicable to the context of the question.

Question 42 (4 marks)

a 20 cm : 200 m

20 cm : 20 000 cm

1 : 1000 ✓

b 12 cm width on diagram represents 12 000 cm
= 120 m

20 cm length on diagram represents 200 m from part **a**.

Area of roof = 200 m × 120 m
= 24 000 m^2 ✓

1 m^3 = 1 kL

So 1800 kL = 1800 m^3

$V = Ah$

$1800 = 24 000h$ ✓

$h = \dfrac{1800}{24\ 000}$

$= 0.075$ m

$= 75$ mm

So 75 mm of rain must fall to collect 1800 kL of water. ✓

> Rainfall questions are commonly asked in HSC exam papers. The assumption is that rainfall is evenly spread through the catchment area (in this case the roof). Remember that rain falling on a roof is basically referring to the volume of a prism, where the height of the prism is the amount of rainfall, in mm. Make sure that all measurements are converted to the same units. Usually, the question will look at rain falling into a rainwater tank and work through the volume and capacity of the tank. Be prepared for these types of questions as well.

The 2020 Mathematics Standard 2 HSC Exam Worked Solutions

The 2020 HSC exam and other past HSC papers can be downloaded from the NESA website (www.educationstandards.nsw.edu.au) by selecting 'Year 11 – Year 12', 'HSC exam papers'. NESA marking feedback and guidelines can also be found there.

WORKED SOLUTIONS

Section I (1 mark each)

Question 1

C The graph of $y = -x^2 + 1$ is a parabola.

> Straightforward question. It is important to know the shapes of non-linear functions and how to manipulate them.

Question 2

D $0.002\,073 \approx 0.0021 = 2.1 \times 10^{-3}$

> Common question to combine scientific (or standard) notation with significant figures. Remember that the decimal point is always moved to follow the first non-zero digit.

Question 3

C $S = \dfrac{D}{T}$

$T = \dfrac{D}{S}$

$= \dfrac{75}{50}$

$= 1.5\,\text{h}$

$= 1\,\text{h}\,30\,\text{min}$

> The distance–speed–time triangle would help answer this type of question.

Question 4

B $FV = PV(1 + r)^n$

$PV = 200, r = \dfrac{0.03}{12} = 0.0025$ per month

$n = 1.5 \times 12 = 18$ months

$FV = \$200(1 + 0.0025)^{18}$

$= \$209.1938\ldots$

$\approx \$209.19$

> Straightforward compound interest question. Remember that the terms 'future value' (FV) and 'present value' (PV) can be applied to the formula.

Question 5

A Absolute error $= \dfrac{1}{2} \times 0.1\,\text{cm} = 0.05\,\text{cm}$

Percentage error $= \dfrac{0.05}{16} \times 100\% = 0.3125\%$

> Common question. Remember that you must find the precision and the absolute error before finding the percentage error.

Question 6

D Let $x = 1$ and $x = 6$ to see the change in the value of y.

When $x = 1$: $y = -1 - 2 \times 1$
$= -3$

When $x = 6$: $y = -1 - 2 \times 6$
$= -13$

Therefore, the value of y decreases by 10 when the value of x increases by 5.

OR The gradient of the linear equation is -2.

If x increases by 5, y increases by $-2 \times 5 = -10$, which is a decrease of 10.

> The best way to approach a question like this is to substitute x-values into the equation to see the changes in the y-values.

Question 7

A 'Positively skewed' refers to the tail of the distribution pointing towards the higher numbers.

> The best way to see the shape is to draw a curve around the peaks of the data.

Question 8

A $z_{\text{French}} = \dfrac{82 - 70}{8}$

$= 1.5$

$z_{\text{Commerce}} = \dfrac{80 - 65}{8}$

$= 3$

$z_{\text{Music}} = \dfrac{74 - 50}{12}$

$= 2$

Therefore, the strongest result was for Commerce and the weakest result was for French.

> z-scores are the best way to compare performance. The higher the z-score, the better the performance compared to the class results.

Question 9

B Each person on Team A needs to play each person in Team B and vice-versa. That means each vertex must connect to 3 other vertices as shown in the diagram.

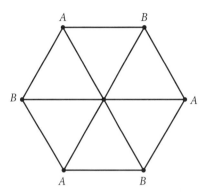

This is a different type of network question.

Question 10

D $C = mt + c$

$c = 90$ (callout fee)

$2 per minute = 60 × \$2 per hour (working rate)$
$$= \$120 \text{ per hour}$$

So $m = 120$

So $C = 120t + 90$, or
$$C = 90 + 120t$$

Applying cost functions to real-life situations is an important skill to practise as it is a common HSC question. Remember the fixed cost is C and the variable cost is m in the first equation.

Question 11

C $S = V_0(1 - r)^n$

$V_0 = 10\,000$, $r = 0.08$, $n = 2 \times 5 = 10$ half-years

$S = \$10\,000(1 - 0.08)^{10}$
$$= \$4343.8845\ldots$$
$$\approx \$4343.88$$

One of the easier depreciation questions. Simple substitution should lead to the correct answer.

Question 12

D A coefficient of -1 means the points are in a perfect straight line and are sloping downwards.

Question 13

A Difference between the BACs $= 0.083\,75 - 0.05$
$$= 0.033\,75$$

So time to reach 0.05 from 0.083 75
$$= \frac{0.033\,75}{0.015}\text{h}$$
$$= 2.25\,\text{h}$$

$10{:}30\,\text{pm} + 2.25\,\text{h} = 10{:}30\,\text{pm} + 2\,\text{h}\,15\,\text{min}$
$$= 12{:}45\,\text{am next day}$$

This BAC question was made harder with the phrase 'expect his BAC to be 0.05', even though the formula is for a person to reach 0 BAC. The difference between the BAC at 10:30 pm and 0.05 is essential to finding the time. Always make sure that you have answered the question.

Question 14

A present value $< \$10\,000 <$ future value

The future value of the annuity is always going to be larger than the present value and the sum of the contributions. The present value is based on the growth of a single lump sum investment at the beginning so it will be the smallest amount as it still needs time to grow.

Question 15

B

	Die					
×	1	2	3	4	5	6
1	1	2	3	4	5	6
2	2	4	6	8	10	12
3	3	6	9	12	15	18
Table section 4	4	8	12	16	20	24
5	5	10	15	20	25	30
6	6	12	18	24	30	36
7	7	14	21	28	35	42
8	8	16	24	32	40	48

There are 48 possibilities and only 4 of those have the result as 6.

$P(6) = \dfrac{4}{48}$
$$= \dfrac{1}{12}$$

For a probability question concerning items with more than 2 options, a table is the best way to see all outcomes in a single space. Set up the table and highlight the results that fit the desired outcomes.

Section II ($\checkmark$ = 1 mark)

Question 16 (4 marks)

a $\tan\theta = \dfrac{8}{10}$ $\checkmark$

$\theta = 39°$ $\checkmark$

b $x^2 = 8^2 + 10^2$ $\checkmark$

$x = \sqrt{164}$

$= 12.8$ $\checkmark$

> A simple question from Years 8–9 maths. As soon as a triangle has a right angle, think of Pythagoras' theorem and trigonometry as your options. Pythagoras' theorem is good to find an unknown side when given the other 2 sides, and SOH CAH TOA is used when an angle is involved. The sine and cosine rules could be used, but these are usually applied to *non-right-angled triangles*, and when used on right-angled triangles, they make the calculations more complicated.

Question 17 (2 marks)

Area of block = $1000 \times 1000 = 1\,000\,000\,\text{m}^2$
Area of section = $5 \times 5 = 25\,\text{m}^2$

$\dfrac{\text{Estimated number of trees}}{1\,000\,000} = \dfrac{8}{25}$ $\checkmark$

Required number of trees = $\dfrac{8}{25} \times 1\,000\,000$

$= 320\,000$ $\checkmark$

> This question compares proportions or ratios. If 8 trees were on a 25 m² block of land, then find out how many trees there are on 1 000 000 m² by equal proportions. Always try to approach these types of questions in a logical way and set out your working clearly.

Question 18 (4 marks)

a Using Prim's or Kruskal's algorithms, the minimum spanning tree could be any one of these:

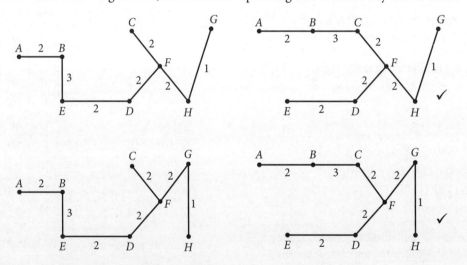

The length of a minimum spanning tree is 14. $\checkmark$

b Adding point K means choosing between KA (12) and KC (10). Therefore, picking the smallest weight and adding to the minimum length: $14 + 10 = 24$ $\checkmark$

> In this question, different answers are possible. Remember that you are trying to find the smallest total to join all vertices without forming any cycles. When drawing the minimum spanning tree, remember to:
> - include all edges
> - label all vertices
> - write all edge weights
> - add up the correct total for the spanning tree.
> This question was well done by most students. Students lost marks for messy diagrams with little detail or labels.

Question 19 (4 marks)

a Drawing a line from $30\,000\,\text{m}^2$, two values result: $x = 100$ and $x = 200$.

The higher value is $200\,\text{m}^2$. ✓

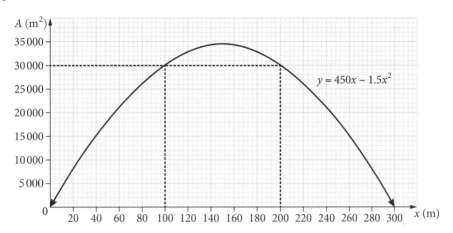

b Maximum occurs when $x = 150$ ✓

$$3x + 2y = 900$$

So $3(150) + 2y = 900$

$$2y = 900 - 450$$

$$y = \frac{450}{2}$$

$$= 225 ✓$$

So $x = 150$, $y = 225$

c Largest possible area $= 150 \times 225$

$$= 33\,750\,\text{m}^2 ✓$$

Questions about reading from a non-linear graph have been asked numerous times in HSC exams. Always think about the context of the graph and understand what each axis represents – questions such as this apply mathematics to real-world situations. The best students know exactly what to do with these problems. This question also focuses on algebraic equation skills, which are important to practise.

Question 20 (3 marks)

Tax payable $= \$20\,797 + 0.37 \times (\$122\,680 - \$90\,000)$

$$= \$32\,888.60 ✓$$

PAYG tax $= \$3000 \times 12$

$$= \$36\,000 ✓$$

Tax refund $= \$36\,000 - \$32\,888.60$

$$= \$3111.40 ✓$$

This is a straightforward and common tax question that was done well. It is important to understand financial terms such as 'PAYG', 'taxable income' and 'tax refund', as well as how to read and calculate from the correct tax/income bracket. Speaking of brackets, when calculating tax payable, remember to use the bracket keys on your calculator to get the right answer. Make sure you answer the 3-mark question completely, and calculate the tax refund.

Question 21 (2 marks)

$FV = PV(1 + r)^n$,

where $FV = 2020$ cost, $PV = 2019$ cost

$FV = 122$, $r = 0.02$ (2% increase), $n = 1$ year

$\$122 = PV(1 + 0.02)^1$

$$= PV(1.02) ✓$$

$PV = \$122 \div 1.02$

$$= \$119.6078\ldots$$

$$\approx \$119.61$$

2019 cost was $\$119.61$. ✓

This is a standard inflation question seeking a value for PV. These questions can ask students to find any of the variables in the PV formula.

Question 22 (3 marks)

Total amount paid $= 500\left(1 + \dfrac{0.17}{365}\right)^{15}$ ✓

$\qquad\qquad\qquad = \$503.504\,56\ldots$ ✓

Balance after \$250 payment $= \$503.504 - \$250$

$\qquad\qquad\qquad\qquad\qquad = \$253.50$ ✓

> 3-mark credit card question with no parts. It is important to make sure that the interest rate is converted to a daily rate. Remember to divide by 365, and notice that credit card accounts charge compound interest, not simple interest. The twist in this question is that the person did not pay off the full amount. Always reread your answer to make sure you have answered the question.

Question 23 (5 marks)

a   1 part pineapple juice $= \dfrac{3}{15}$ ✓

$\qquad$ 4 parts orange juice $= 4 \times \dfrac{3}{15}$

$\qquad\qquad\qquad\qquad\quad = 0.8\,\text{L or } 800\,\text{mL}$ ✓

b   Volume $= 40 \times 35 \times 20$ ✓

$\qquad\qquad\quad = 28\,000\,\text{cm}^3$

$\qquad$ 1 cm^3 = 1 mL

$\qquad$ Capacity = 28 000 mL

$\qquad\qquad\qquad = 28\,\text{L}$ ✓

$\qquad$ Total number of parts $= 15 + 9 + 4$

$\qquad\qquad\qquad\qquad\qquad\quad = 28$

$\qquad$ 28 parts = 28 L

$\qquad$ 1 part = 1 L

$\qquad$ So 9 parts = 9 L of mango juice required. ✓

> Spend some time planning your solution before writing. Some students wrote answers that were more complicated than needed. Some did not check whether their answers sounded reasonable. When dealing with ratios, identify all the given information and find out what quantity 1 part equals. The conversion between volume and capacity is an important concept and is used in many ways in HSC questions.

Question 24 (4 marks)

a and **b**

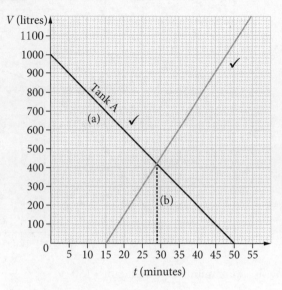

From the graph, the value of t when 2 tanks contain the same volume of water is 29 minutes. ✓

c

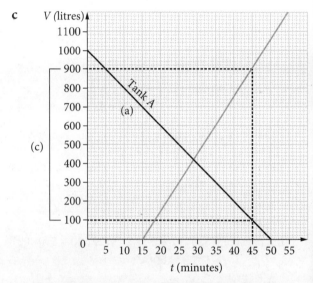

At 45 minutes, Tank A is at 100 L and Tank B is at 900 L. ✓

> This is a 4-mark linear modelling question. Drawing graphs from a linear function or a worded description is an underrated skill that needs to be mastered. A simple table of values will help with this. The point of intersection of two linear functions is the solution to the 2 equations. Always be neat and use a ruler when drawing these graphs. This will help when reading from the graph is required.

WORKED SOLUTIONS

Question 25 (3 marks)

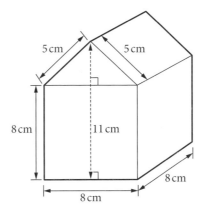

Height of triangle: $11 - 8 = 3$

Surface area = 2 × triangle faces + 5 square faces (walls and bottom) + 2 × rectangular faces (roof)

$$= 2 \times \left(\frac{1}{2} \times 8 \times 3\right) + 5 \times (8 \times 8) + 2 \times (5 \times 8)$$

$$= 424 \, \text{cm}^2 \checkmark$$

A 3-mark surface area question with no parts/clues where the only tricky part is finding the perpendicular height of the triangle before calculating the areas of all the faces. Be systematic and logical in your setting out – this is a question where students are prone to making small errors.

Question 26 (5 marks)

a   The minimum completion time is 46 minutes. ✓

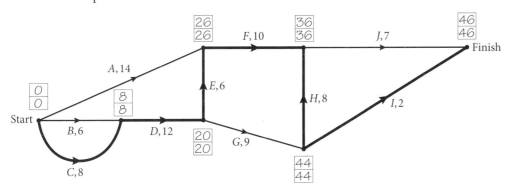

b   The critical path is *CDEFHI*. ✓✓

c

Activity	Earliest start time	Float time	
A	0	26 − 0 − 14 = 12	✓
G	20	44 − 20 − 9 = 15	✓

Finding the critical path of a network is a very important skill to master. Being able to do a forward and backward scan quickly and accurately will be advantageous. When calculating float time, do not just subtract the EST from the LST. Always calculate float time as LST of next activity − EST of activity in question − duration of activity in question. This will result in the correct float time always.

Question 27 (5 marks)

a $A = \frac{h}{2}\left(d_f + d_l\right)$

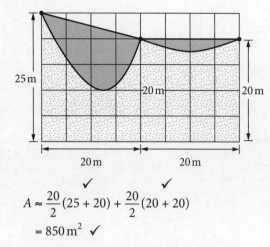

$A \approx \frac{20}{2}(25 + 20) + \frac{20}{2}(20 + 20)$

$= 850\,\text{m}^2$ ✓

b Estimate will be more than the actual area ✓
as the trapeziums include areas that are not
part of the garden (see above diagram). ✓

> The trapezoidal rule is a simple concept and is
> important to get right as these are marks that
> most students will achieve. Practise reading a
> scale diagram and applying the trapezoidal rule
> twice for a problem. Remember that the result
> is only an estimate of the actual area.

Question 28 (4 marks)

Mean of first data set = 8 ✓

Median of first data set = 9 ✓

Median of second data set = 9.5 ✓

Mean of second data set = $\frac{40 + x}{6}$

$\frac{40 + x}{6} - 8 = 5$

$\frac{40 + x}{6} = 13$

$40 + x = 78$

$x = 38$ ✓

> The HSC students struggled with this question.
> When approaching a complex 4-mark, no-clues
> question like this, start by chipping away at
> the easier concepts. Find the mean of the first
> data set and median of both data sets. That will
> then give some clues to get to the final answer.
> Forming that equation was crucial. A common
> mistake is to ignore the basics and try to tackle
> the harder concept first. Always show what you
> know rather than focusing on what you don't.

Question 29 (3 marks)

Dividend from Company $ABC = 200 \times \$5.50 \times 0.06$
$ = \$66$ ✓

Dividend from $XYZ = \$149.52 - \$66$
$ = \$83.52$ ✓

Number of shares in $XYZ \times 0.04 \times 6 = \$83.52$

Number of shares in $XYZ = \frac{83.52}{0.24}$
$ = 348$ ✓

> Another 3-mark question with no parts. The best
> students took a structured approach and set
> out their work clearly. A formula for calculating
> dividend yield is not on the HSC reference sheet.
> It is important to understand what it is and walk
> through the bigger picture of shares and how
> they work. Understand the financial terms and
> definitions, and know what a dividend is.

Question 30 (3 marks)

a

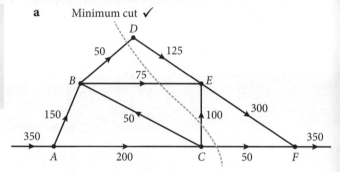

Maximum flow = minimum cut
$ = 50 + 75 + 100 + 50$
$ = 275$ ✓

b

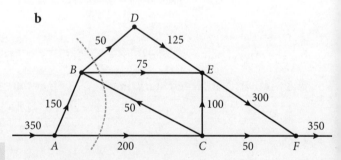

Value of this cut = 50 + 75 + 200
$ = 325$ ✓

This is not equal to the maximum flow,
therefore it is not the minimum cut.

> When performing cuts, it may be helpful to look
> for the smallest edges and see if a cut could go
> through those to form the minimum. Remember
> that arrows pointing back towards the source
> (CB) are not counted on the cut. The minimum cut
> represents the maximum flow of the network, and
> this is the simplest method of solving part **a**.

Question 31 (5 marks)

a $\angle APB = 100° - 35°$ (difference between
       $= 65°$ ✓   the bearings)

b

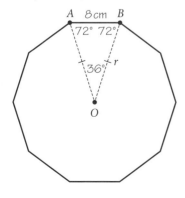

$$AB^2 = 7^2 + 9^2 - 2 \times 7 \times 9 \cos 65° \checkmark$$
$$= 76.7500\ldots$$
$$AB = 8.76007\ldots$$
$$\approx 8.76 \, \text{km} \checkmark$$

c Need to find $\angle PAB$ in order to find the
bearing of B from A.

$$\frac{\sin A}{9} = \frac{\sin 65°}{8.76\ldots}$$
$$\sin A = \frac{9 \sin 65°}{8.76\ldots}$$
$$A \approx 69° \checkmark$$

$$\text{Bearing} = 180° - (69° - 35°)$$
$$= 146° \checkmark$$

This question is for students aiming for Band
5 or 6. Bearings questions combined with the
sine or cosine rule have typically been some of
the harder questions in the HSC, with the 'show
that' answer to part **a** being used to solve part **b**,
and so on. It is important to practise bearings
and look for keywords like 'from' to help draw
diagrams correctly. Use alternate angles between
parallel north lines to find angles, and check
whether answers look reasonable.

Question 32 (4 marks)

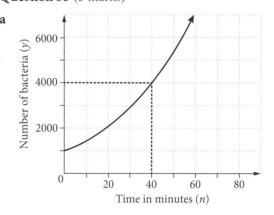

$$AB = 80 \div 10$$
$$= 8 \, \text{cm} \checkmark$$

$$\angle AOB = \frac{360°}{10}$$
$$= 36°$$

$$\angle OAB = \frac{180° - 36°}{2}$$
$$= 72° \checkmark$$

$$\frac{r}{\sin 72°} = \frac{8}{\sin 36°}$$
$$r = 12.94\ldots$$

$$\text{Area} = \frac{1}{2} \times 12.94\ldots \times 12.944\ldots \times \sin 36°$$
$$= 49.2429\ldots \checkmark$$

$$\text{Total area} = 49.2429\ldots \times 10$$
$$\approx 492.4 \, \text{cm}^2 \checkmark$$

This 4-mark problem with no parts/clues relied
heavily on knowledge of geometrical concepts
such as isosceles triangles, regular decagons
and interior angles. This was a harder question
that required students to analyse the information
correctly and do some thinking and planning
before calculating using the sine rule and the
area of a triangle formula. Even though simple
geometrical concepts are not in the HSC course,
they are assumed knowledge from Years 9 and 10
and it would be wise to be familiar with these.

Question 33 (3 marks)

a

Number of bacteria in 40 minutes
$= 4000$ bacteria ✓

b $y = A \times b^n$

A is vertical intercept, so $A = 1000$.

$y = 1000b^n$

When $n = 20$, $y = 2000$.

$2000 = 1000 \times b^{20}$ ✓

Using trial and error:

$1000 \times (1.03)^{20} = 1806.11$

$1000 \times (1.03)^{20} = 2191.12$

So lower estimate = 1.03

and upper estimate = 1.04 ✓

> Reading from a non-linear graph is a common type of question. It is important to understand the initial value of the exponential curve (the y-intercept, or where $x = 0$). It could be read from the y-intercept on the graph. The guess-and-check method is the recommended method in Mathematics Standard 2 for finding the power, n. This is a concept that is worth practising.

Question 34 (4 marks)

a $A_1 = 60\,000(1.05) - 800$

$= \$59\,500$ ✓

$A_2 = \$59\,500(1.005) - 800$

$= \$58\,997.50$

$A_3 = \$58\,997.50(1.005) - 800$

$= \$58\,492.49$ ✓

b Total withdrawals after 3 months = 800×3

$= \$2400$

Balance reduced by
$\$60\,000 - \$58\,494.49 = \$1507.51$ ✓

Interest = $\$2400 - \$1507.51$

$= \$892.49$ ✓

> Even though the question was complex and unfamiliar, algebraic substitution can be used to find the values of the first 3 terms. The value of A_1 indicates that $n = 1$, which would then be applied to the original formula. The recurrence formula is built on top of the previous equation, so it was important to calculate the equation 3 times and use the answers from the previous term to achieve the final result. Note that part **a** leads to part **b**.

Question 35 (7 marks)

a

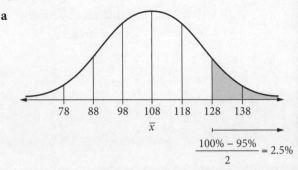

$$\frac{100\% - 95\%}{2} = 2.5\%$$

128 corresponds to 2 standard deviations above the mean.

Therefore 2.5% have an IQ score higher than Yin's. ✓

b

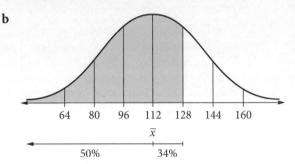

Yin's score is 1 standard deviation above the mean. Therefore, the percentage of the population that have an IQ score less than

128 is $50\% + \frac{68\%}{2} = 84\%$. ✓

So expected number of adults = $84\% \times 1\,000\,000$

$= 840\,000$ ✓

c Let x be Simon's IQ score:

z-score in City A $= \dfrac{x - 108}{10}$ ✓

z-score in City B $= \dfrac{x - 112}{16}$

$\dfrac{x - 108}{10} = \dfrac{x - 112}{16}$

$16(x - 108) = 10(x - 112)$ ✓

$16x - 1728 = 10x - 1120$

$6x = 608$

$x = \dfrac{608}{6}$

$= 101.333\ldots$

≈ 101.3 ✓

> This question combined z-scores with forming and solving equations. The empirical rule from the HSC reference sheet is a common application of z-scores. Note that the percentages can be used in probability problems such as expected value.

Question 36 (5 marks)

From box plot, median temperature = 22. ✓

So mean temperature, $\bar{x} = 22 - 0.525 = 21.475$ ✓

Mean chirps, $\bar{y} = \dfrac{684}{20} = 34.2$ ✓

Since $(\bar{x}, \bar{y})$ is on the line:

$$y = -10.6063 + bx$$
$$34.2 = -10.6063 + b(21.475)$$
$$b = \frac{44.8063}{21.475}$$
$$= 2.0864\ldots \checkmark$$

So $y = -10.6063 + 2.0864\ldots x$

When $x = 19$:
$$y = -10.6063 + 2.0864\ldots \times 19$$
$$= 29.0360$$
$$\approx 29 \text{ chirps} \checkmark$$

This is the famous 'crickets' question that caused the most concern in the 2020 Standard 2 paper. It was the first 5-mark question that was a common question to both the Standard and Advanced exams, which required a lot of reading and unpacking of a list of clues. The difficulty was that the mean of the data had to be assumed to be a sample mean. This resulted in a substitution into the equation of the line given. For a question such as this, always work through the smaller concepts first, such as mean and median. Piece those concepts together and work through the question logically and systematically. Do not round off in the middle of the problem, but at the very end.

Question 37 (3 marks)

Look at two amounts: A and B.

A: 20 years, $1000 per year

B: 20 years, 10 years

For A:

$$0.02 \times 20 \text{ periods} = 16.351$$
$$1000 \times 16.351 = \$16\,351 \checkmark$$

For B:

Need \$30 000 for the last 10 years of B to be able to pay \$3000 each year for 10 years.

So how much is needed to invest at the beginning of the 10-year period?

$$\$3000 \times 8.983 = \$26\,949 \checkmark$$

What amount is needed at the beginning of the loan to get to \$26 949?

$$FV = PV(1 + r)^n$$
$$\$26\,949 = PV(1 + 0.02)^{20}$$
$$\$26\,949 = PV(1.02)^{20}$$
$$PV = \frac{\$26\,949}{(1.02)^{20}}$$
$$= \$18\,135.9044\ldots$$
$$\approx \$18\,135.90$$

So total that needs to be invested
$$= \$16\,351 + \$18\,135.90$$
$$= \$34\,486.90 \checkmark$$

This challenging final question again had no parts and was aimed at students hoping to achieve a Band 6 as it tested deep understanding of annuities. This question was wordy and difficult from a conceptual standpoint. Understanding the amount needed at the end of 20 years so that \$3000 could be drawn upon each year was the fundamental piece of the puzzle. Work through this question a few times to make sure that you understand the entire context of the question.

HSC exam reference sheet

Mathematics Standard 1 and 2

© NSW Education Standards Authority

Measurement

Limits of accuracy

Absolute error $= \frac{1}{2} \times$ precision

Upper bound = measurement + absolute error

Lower bound = measurement − absolute error

Length

$$l = \frac{\theta}{360} \times 2\pi r$$

Area

$$A = \frac{\theta}{360} \times \pi r^2$$

$$A = \frac{h}{2}(a + b)$$

$$A \approx \frac{h}{2}(d_f + d_l)$$

Surface area

$$A = 2\pi r^2 + 2\pi rh$$

$$A = 4\pi r^2$$

Volume

$$V = \frac{1}{3}Ah$$

$$V = \frac{4}{3}\pi r^3$$

Trigonometry

$$\sin A = \frac{\text{opp}}{\text{hyp}}, \quad \cos A = \frac{\text{adj}}{\text{hyp}}, \quad \tan A = \frac{\text{opp}}{\text{adj}}$$

$$A = \frac{1}{2}ab\sin C$$

$$\frac{a}{\sin A} = \frac{b}{\sin B} = \frac{c}{\sin C}$$

$$c^2 = a^2 + b^2 - 2ab\cos C$$

$$\cos C = \frac{a^2 + b^2 - c^2}{2ab}$$

Financial Mathematics

$$FV = PV(1 + r)^n$$

Straight-line method of depreciation

$$S = V_0 - Dn$$

Declining-balance method of depreciation

$$S = V_0(1 - r)^n$$

Statistical Analysis

An outlier is a score

less than $Q_1 - 1.5 \times \text{IQR}$

or

more than $Q_3 + 1.5 \times \text{IQR}$

$$z = \frac{x - \mu}{\sigma}$$

Normal distribution

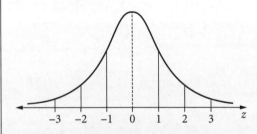

- approximately 68% of scores have z-scores between −1 and 1

- approximately 95% of scores have z-scores between −2 and 2

- approximately 99.7% of scores have z-scores between −3 and 3